A COURSE IN WORLD GEOGRAPHY—YOUNG & LOWRY

Book

EUROPE
and The Soviet Union

By

J. H. LOWRY, M.A., B.Sc. (Econ.)

Senior Geography Master
Cranleigh School

Second Edition

MAPS AND DIAGRAMS BY J. H. LOWRY

LONDON
EDWARD ARNOLD (PUBLISHERS) LTD

© E. W. Young and J. H. Lowry 1972

First published 1966
by Edward Arnold (Publishers) Ltd.,
25 Hill Street, London W1X 8LL
Reprinted 1966
Reprinted 1967
Reprinted 1968
Reprinted 1969
Reprinted 1970
Reprinted 1971
Second Edition 1972

ISBN: 0 7131 1715 x

The full course comprises:

BOOK 1—People in Britain
BOOK 2—People Round the World
BOOK 3—Regions of the World
BOOK 4—The British Isles
BOOK 5—The World: Physical and Human
BOOK 6—Europe and the Soviet Union
BOOK 7—North America
BOOK 8—East Africa
BOOK 9—Central Africa

Printed by offset in Great Britain by
William Clowes & Sons, Limited, London, Beccles and Colchester

FOREWORD

THIS book forms one of the regional studies in a series designed primarily for pupils of moderate ability who will take geography at G.C.E. 'O' Level. First-year sixth form and training college students will also find this volume useful as a basis for more advanced work, for it contains a very large number of illustrations and sketch maps nòt readily available elsewhere.

In keeping with other books in the series, formal statement of geographical fact has been cut to a minimum and the necessary data are expressed very largely in the form of maps, diagrams and illustrations. The responsibility for their interpretation is laid expressly on the pupil, with or without the teacher's help according to circumstances. In this way it is hoped to encourage the development of genuine geographical understanding and interest, and to discourage mere memorisation. The layout of every page has been carefully devised so that, although each book is written in chapter form, the material does in fact fall into units of one, two or at most four pages. Lesson-planning will be found to be much facilitated by this arrangement.

Books I and II are by E. W. Young and Books III and IV by J. H. Lowry in association with E. W. Young, who has planned the series as a whole. The fifth-year books are the work of both authors, who between them have taught geography as a G.C.E. subject in secondary modern, grammar and public schools and have served on various G.C.E. examining boards.

Chapter 1 of Book 6 offers a general survey of European structure, relief, climate, natural vegetation, economic resources and political units: pupils are referred back to this basic information throughout the remainder of the book. The regional chapters fall into five main groups as follows: *Northern Europe*: Norway, Sweden, Finland, Denmark and Iceland (Ch. 2); *Western and Central Europe*: France, West Germany, the Low Countries and Luxembourg, Switzerland and Austria (Chs. 3–6); *Southern Europe*: Italy, Spain, Portugal and Greece (Chs. 7 and 8); *Eastern Europe*: East Germany, Poland, Czechoslovakia and the Balkans (Ch. 10) and *The Soviet Union* (Ch. 9). This arrangement takes account both of political and of economic realities. West and East Germany, for example, are dealt with separately, and the economic links between the Soviet Union and Eastern Europe are readily apparent. Both the European and Asiatic regions of the

Soviet Union are considered in Chapter 9. In deciding on this course the author had two points in mind, viz. (i) the U.S.S.R. is now so well integrated that division at the Urals is arbitrary and geographically unreal and (ii) there is a dearth of elementary textbooks on this major world power.

In preparing this book every effort has been made to keep abreast of and give due emphasis to contemporary economic and political changes. Such schemes as the Southern Italy Development Fund, the Midi Irrigation Project, the Virgin Lands Programme and collectivisation in S. E. Europe are rapidly changing the physical and human landscapes over very large areas. The author acknowledges with thanks the information relating to these and other projects readily given to him by the embassies and trade representatives of the countries involved.

FOREWORD TO SECOND EDITION

IN THE five years since this book was first published considerable changes have occurred in European geography. In preparing this enlarged Second Edition opportunity has been taken both to revise the existing text and to incorporate new material. The main additions comprise sixteen new pages in the chapters on West Germany (economic changes in the Ruhr, Rhineland and Hamburg regions); the Netherlands (development of Europort, Randstad Holland and the Groningen gas-field); Italy (economic and social reconstruction of the South) and Eastern Europe (fuller details of recent industrial and agricultural trends). Many other amendments have been made throughout the book and statistical information has been brought up to date and metricated.

J.H.L.

CONTENTS

SPECIAL NOTE

Throughout this book certain words are printed in heavy type, thus: **escarpment**. It is suggested that these words should be collected, chapter by chapter; that a satisfactory meaning should be attached to them; and that the words and their definitions should be recorded in the pupil's own 'geography dictionary' or in any other form that suits the teacher's particular approach. By this means a sharper edge may perhaps be left on terms which all too often are blurred by generalisation.

CHAPTER 1
Europe as a Whole

R*ELIEF AND STRUCTURE.* Look at an atlas relief map of Europe. Notice that in the east the land lies mostly level and unbroken—the plains of Northern Germany and Poland widening and gradually merging into the steppes of Russia and the Ukraine (*photo D, p. 23*). Farther west and south, by contrast, the terrain is remarkably varied; high mountain ranges, plains just above sea level, river valleys and basins; all lie close to one another with no obvious pattern (*photo opposite, and pp. 20–25.*) In the west the waters of the Atlantic Ocean and its marginal seas have invaded and divided the land, giving it an irregular shape and a lengthy coastline.

Europe has been called 'a peninsula of peninsulas'. (*1*) What does this mean? (*2*) On a bold sketch-map name the peninsulas of Europe.

The physical contrast between Eastern Europe and the rest of the continent results from differences in their geological histories. The rocks which in Eastern Europe form the **Russian Platform** (*map overleaf*) have lain virtually undisturbed for hundreds of millions of years, whereas those of Western, Central and Southern Europe have suffered the upheavals and dislocations of many violent 'earth-storms'. The oldest part of Europe is the **Baltic Shield** (*V: 26 and 35*). This region forms the geological 'heart' of the Continent. Formerly it was probably a mountainous region, but today it is a low, undulating peneplain, having suffered over 500 million years of denudation; its rocks are hard, metamorphic and crystalline. Since the Baltic Shield was formed fantastic changes have occurred. Arid deserts; tropical sea-coasts; swampy mud-flats; towering mountain chains—all these and many other landscapes have emerged and then gradually been worn away. Each has added something to the present-day geological pattern of Europe, which can best be understood in relation to three great periods of mountain building:—

I. About 320 million years ago, ranges of folded mountains extending into what is now the northern part of the British Isles were thrust up against the north-western edge of the Baltic Shield. Although much altered by long-continued denudation, relics of these mountains survive as worn-down plateaus (*photo B, p. 21* and V: 30).

6

The Zillertaler Alps, Austria. Young fold mountains such as these are subject to active denudation. Note the shrunken glacier.

Their geological name '**Caledonian**' suggests that the effects of this 'earth-storm' were first studied—(*3*) *where?*

2. About 240 million years ago another period of mountain building caused the upheaval of the bed of a great sea extending from east to west right across Central Europe. Included in these sediments were vegetable remains of swamp forests which in time were gradually compressed into coal. Shattered, re-elevated remnants of the folded mountain ranges formed at this time still survive (*photo G, p. 25*). The most important of them are indicated on the map overleaf by the letter H. (*4*) Name these ranges from your atlas and from the list on the map. Notice that one such area is the Harz Mountains in Germany; the Latin name for these, *Hercynia*, is used to denote the whole **Hercynian** mountain system. Most of Europe's important coal fields lie along the northern foothills of, or within, these uplands.

(*contd. on p. 10*)

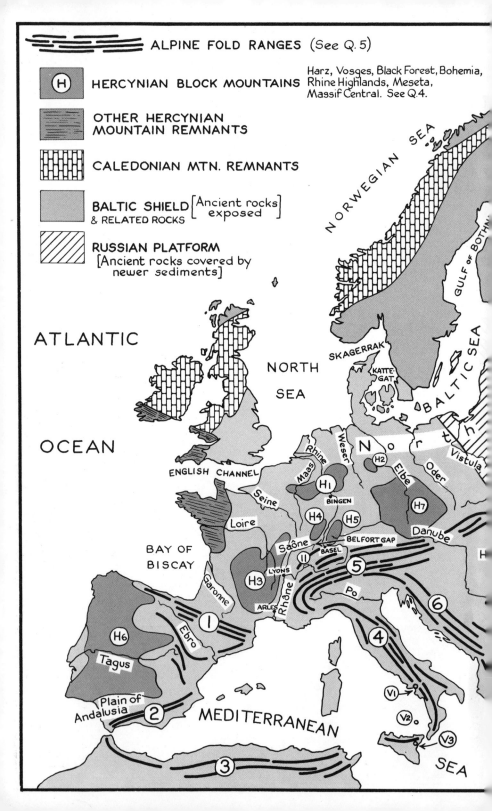

ALPINE FOLD RANGES (See Q.5)

H HERCYNIAN BLOCK MOUNTAINS
Harz, Vosges, Black Forest, Bohemia, Rhine Highlands, Meseta, Massif Central. See Q.4.

OTHER HERCYNIAN MOUNTAIN REMNANTS

CALEDONIAN MTN. REMNANTS

BALTIC SHIELD & RELATED ROCKS [Ancient rocks exposed]

RUSSIAN PLATFORM [Ancient rocks covered by newer sediments]

ATLANTIC

OCEAN

NORWEGIAN SEA

GULF OF BOTHNIA

BALTIC SEA

NORTH SEA

SKAGERRAK

KATTE-GAT

N o r t h

ATLANTIC

ENGLISH CHANNEL

Seine

Loire

BAY OF BISCAY

Rhine

Maas

Weser

H2

H1

BINGEN

H4   H5

Saône

Elbe

Oder

Vistula

H7

Danube

BELFORT GAP

BASEL

5

LYONS

H3

ARLES   Rhône

II

Po

6

Garonne

Ebro

1

4

H6

Tagus

Plain of Andalusia

2

MEDITERRANEAN

V1

V2

V3

SEA

3

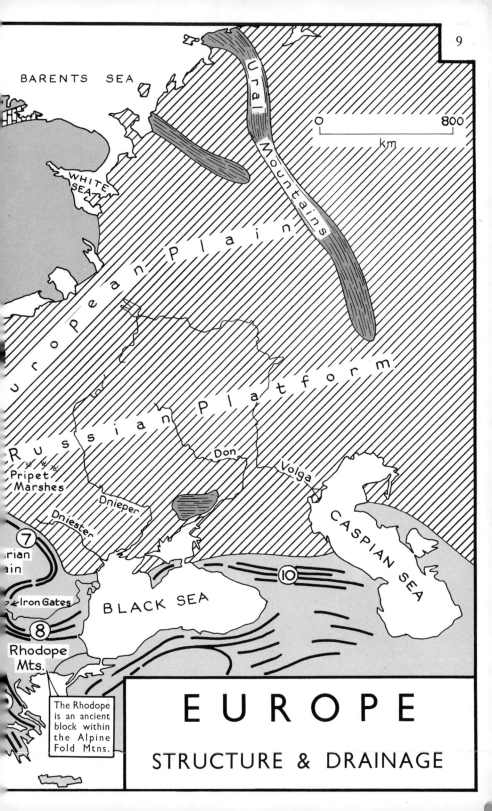

BARENTS SEA

WHITE SEA

Ural Mountains

0   800
km

E u r o p e a n   P l a i n

R u s s i a n   P l a t f o r m

Don

Volga

Pripet
Marshes

Dnieper

Dniester

CASPIAN SEA

⑦

rian
ain

Iron Gates

⑧

Rhodope
Mts.

BLACK SEA

⑩

The Rhodope
is an ancient
block within
the Alpine
Fold Mtns.

E U R O P E

STRUCTURE & DRAINAGE

3. Much more recently, about 35 million years ago, a further period of mountain-building took place. Yet another series of towering folded mountain ranges were formed, curving across Central and Southern Europe and extending into the north-west tip of Africa. Because, geologically speaking, these mountains are still young, they have been less affected by denudation and so today form the highest and most impressive peaks in Europe (*photos pp. 7, 133 and 152*). The most intensive folding took place in the Alps, and the whole system is therefore known as **Alpine**. The following important Alpine mountain ranges are indicated on the map on page 8 by numbers 1–11:—Alps, Apennines, Atlas Mtns., Sierra Nevada, Pyrenees, Jura, Illyrian Alps, Pindus, Carpathians, Balkans, Caucasus. (5) Write down each name with its appropriate number.

All these periods of mountain building were times of violent earthquakes and volcanic activity. ((*6*) *Why?*) Fissures and vents appeared in the Earth's crust, and molten lava forced its way just below or on to the surface. Volcanoes associated with the Caledonian and Hercynian 'earth-storms' have long since become extinct, but some of those within the Alpine ranges are still active.

Three active volcanoes labelled V_1–V_3 (*but not in that order*) on the map on page 8 are: Vesuvius, Etna, Stromboli. Write down each name with its correct label. Each of them is a *cone* volcano, the characteristic shape of which is seen in the photograph of Vesuvius on page 169.

For 205 million years, between Hercynian and Alpine times, most of Europe except the north was submerged beneath seas or lakes. Widespread deposits of sands, limestone-forming oozes and clay-forming muds, dumped beneath these waters, were raised above sea-level during the Alpine uplift. The less elevated regions are represented today by basins and plains in which the strata remain more or less level, or show gentle warping and tilting. Such places include S.E. England, N. and S.W. France, N. Germany, Hungary and N. Italy. Elsewhere rift valleys were formed when strata collapsed between lines of parallel faults; examples include the Rhine Valley between Basel and Bingen, the Rhône Valley between Lyons and Arles, and the Plain of Andalusia. (7) Name the highlands bordering each of these depressions.

Since 'Alpine' times the natural landscape of Europe has been sculptured mainly by the **Great Ice Age**. About 600 000 years ago the climate gradually became so cold that huge ice-sheets accumulated in various highland regions. The main sheet (*see* *map*), centred on Scandinavia and in places more than 3000 m thick, gradually spread out to cover an area of some 8 million square kilometres. Other smaller ice-sheets were centred on the Alps, the Pyrenees and the uplands of Northern Britain. These ice sheets advanced, and later melted, at least four times as the climate slowly fluctuated between extreme cold and sub-tropical warmth. The most recent retreat probably began about 25 000 years ago. The main effects of these ice-sheets were:—

(*a*) *Erosion on a vast scale, involving the removal of immense quantities of soil and rock fragments from highland areas.* Glaciated mountains in Europe, as elsewhere, are characterised by such features as over-deepened U-shaped valleys, cirques, hanging valleys, arêtes, pyramidal peaks and truncated spurs. The photographs on pp. 7 and 133 include several of these features: (*8*) make large outline sketches of them and add appropriate labels.

(*b*) *Deposition in lowlands of enormous quantities of boulder clay and moraine* (*photo, p. 50*). The main glacial dumping grounds were the plains of Poland and Northern Germany, the Bavarian Plateau and the Po Valley in Italy. ((*9*) For which ice-sheets respectively?) In addition, melt-water flowing from the edges of the ice-sheets carried away loads of sand, gravel and 'rock-flour' (pulverised rock) from the terminal moraines. Hence **glacial outwash deposits** were laid down in immense sheets, especially on the lowlands of Northern Europe.

(*c*) *Formation of loess.* During drier phases of the Ice Age

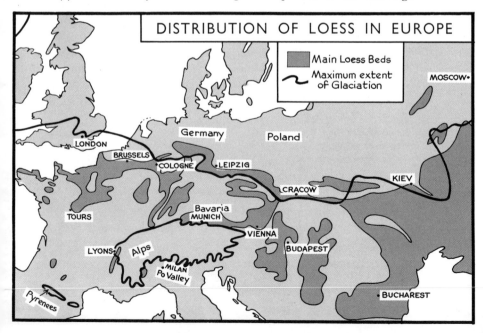

southward-blowing winds raised great clouds of fine dust from moraines marking the edge of the Scandinavian ice-sheet. These particles in time drifted back to earth, blanketing the landscape beneath a layer of dust up to 100 m thick. This deposit is called **loess** in German, **limon** in French and sometimes **brick-earth** in English. Fine-textured, fertile, loamy soils developed on the loess. The map on page 11 shows the principal European loess deposits, many of which run in an east–west belt along the northern flanks of the Hercynian Uplands. Note their location carefully, for the loess districts form most of the principal agricultural areas of Europe (*photo C, p. 22 and diagram, p. 236*).

(*d*) *Fluctuations in sea-level.* As the ice-sheets melted, sufficient water flowed into the oceans to cause a considerable world-wide rise in sea-level. This brought about marked changes in the shape of Europe's coastline. The rising waters joined the Baltic 'lake' by narrow sea-straits to the North Sea, and in about 6 000 B.C. they separated Britain from the Continent. River valleys of upland coasts were flooded, as in N.W. France, to form **rias** (*map, p. 76*) and the glaciated valleys of Norway were converted into **fiords** (*photo, p. 34*). Along low-lying coasts the higher sea-level produced tidal mud-flats, broad shallow lagoons such as the Baltic *haffs* (*p. 246*), and low off-shore islands, e.g. the Frisian Islands (*p. 112*).

Since glacial times the relief of Europe has been modified mainly by river action. River *erosion*, for example, has notched the lips of many **hanging valleys** (*see diagram*), whilst *deposition* of alluvium has changed thousands of square kilometres of former sea-bed into fertile plains and islands. Examples of the latter include the Rhine–Maas delta and the lowlands at the head of the Adriatic Sea. Europe's rivers can be divided into three main groups, according to whether they drain:—(i) north and west to the Atlantic Ocean and its marginal seas; (ii) south to the Mediterranean; (iii) south

Modification of hanging valley by river action.

or east into the Black or Caspian Seas. (*10*) Using an atlas, compile lists to fit each of these groups including the following rivers:—Garonne, Rhône, Loire, Ebro, Dnieper, Seine, Po, Meuse (Maas), Rhine, Dniester, Weser, Elbe, Don, Oder, Vistula, Danube, Volga.

As the relief of much of Europe is so broken by mountains and uplands, these rivers and their fertile basins have always been vitally important to Man. In addition to providing valuable land for agriculture, certain gaps or gateways have been paths of migration and communication since pre-historic times. For example, civilised people first reached N.W. Europe from the Mediterranean via the Rhône–Saône Valley, and this route is still a major avenue of trade and passenger traffic. (*11*) Use your atlas and the map on page 8 to work out the following 'heads and tails':—

The Belfort Gap

The Rhône–Saône Valley

The Iron Gates

The Rhine Valley

} ? {

guard the eastern entrance to the fertile Hungarian Plain.

provides a route linking the Rhineland to the Mediterranean.

is an artery of trade from Rotterdam to Basle.

provides the shortest route from the Mediterranean to the English Channel.

(*12*) Make a list of all the countries through which each of the following rivers runs:—Rhine, Rhône, Danube, Elbe.

Many important European cities, including several capitals, are located on large rivers: (*13*) name five such cities and (*14*) suggest how a riverside location may have fostered their growth. (*15*) Name the rivers on which the following ports are located:—Bordeaux, Rotterdam, Odessa, Antwerp, London, Hamburg, Bremen, Rouen, Lisbon, Nantes. (*16*) Why is the mouth of a river a particularly favourable location for a port?

The Iron Gates hydro-power and shipping project.

CLIMATE. The climate of Europe is in many respects remarkable. Consider for example the following:—

Northern Italy in January is nearly as cold as Iceland.

The Arctic coast of Norway is always ice-free, yet 800 kilometres farther south the Gulf of Bothnia freezes every winter.

In the Isles of Scilly sub-tropical plants grow at all times, yet in the same latitude in the Ukraine the growing season is only 28 weeks. Summers in the Southern Ukraine, however, are nearly as hot as on the Equator.

To understand these peculiarities one must bear in mind the position of Europe, especially in relation to (*a*) the Atlantic Ocean; (*b*) the great Eurasian land-mass; (*c*) the paths taken by 'depressions' and (*d*) the belt of sub-tropical calms and droughts.

Influences of the Atlantic Ocean. That oceanic influences play a very great part in determining European climate conditions can readily be seen from this map. Gibraltar is about latitude 36°N. and the North Cape about latitude 71°N., i.e. well within the Arctic Circle. Because of this position the whole of Europe except the extreme east is affected most of the year by the **Westerlies**. These winds reach Europe from a more or less westerly direction, i.e. from the Atlantic Ocean.

Since the surface waters of the North Atlantic are brought from tropical regions by the Gulf Stream Drift (*map below, and V: 65–69*) they are exceptionally warm for their latitude. Air is heated

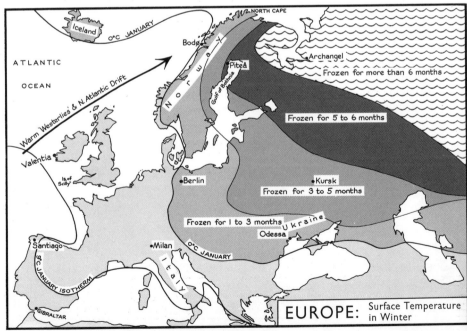

EUROPE: Surface Temperature in Winter

or cooled mainly by the surface—land or water—over which it is blowing; so the Westerlies, too, are very warm for their latitude.

Their effect is particularly noticeable in winter, when the Earth's surface is receiving from the Sun each day less heat than it is losing by radiation into space. The Earth therefore cools, and the land cools faster than the sea (*III: 10–11*). On the western coasts, however, the prevailing onshore winds keep temperatures relatively high in winter and low in summer, as shown in the tables below.

(*17*) Trace a large outline map of Europe, and from the map opposite mark and name all the towns mentioned in the tables. Write beside each town its mean annual temperature range. Compare each west coast town with those lying farther east in the same latitude, in respect of (*a*) summer and (*b*) winter temperature and (*c*) annual range of temperature. State and account for the differences noted.

The relatively mild winters and cool summers give west coast districts of Europe an **equable climate**. It is also a *wet* climate, for the Westerlies are very moist, having absorbed great quantities of water-vapour from the ocean. When the warm, moist air reaches land it is likely to be chilled by being forced up over high ground, or by convectional turbulence (*IV: 54–9*). As a result the water-vapour is condensed and precipitated as rain or snow, especially on seaward slopes.

Western and Central European countries are therefore characterised by moderate to heavy rainfall, with marked '**rain-shadow**' zones. Compare the relief and rainfall maps of Europe in your atlas and (*18*) name three 'rain-shadow' areas and the mountains behind which they lie. (*E.g. Eastern Sweden, in the lee of the Scandinavian Highlands.*)

Influences of the Eurasian land-mass. We have noted that in winter the land loses its heat more rapidly than the sea. The air over the land is therefore chilled, becomes denser and gradually sinks. Thus in winter the Eurasian land-mass becomes a marked high-pressure centre; anticyclones develop and bitterly cold 'continental' air gradually spreads out from the interior. This

NORTH-WEST EUROPE					CENTRAL EUROPE					EASTERN EUROPE			
Place	Mean Temp. °C		Rain-fall mm		Place	Mean Temp. °C		Rain-fall mm		Place	Mean Temp. °C		Rain-fall mm
	Jan.	July				Jan.	July				Jan.	July	
Bodø	−1	13	725		Piteå	−8	15	525		Archangel	−14	16	425
Valentia	7	15	1400		Berlin	−1	18	550		Kursk	−9	20	500
Santiago	7	18	1625		Milan	1	24	1000		Odessa	−3	23	400

explains why Eastern Europe has such very cold winters, for blasts of icy air coming from Arctic latitudes regularly sweep across the plains of Russia in that season. Tongues of this air, pushing west-wards and southwards, also lower temperatures in Central Europe; sometimes they penetrate as far as the British Isles and Mediter-ranean lands (IV:61). Lengthy periods of settled, frosty and sometimes foggy weather then ensue, but although the air is extremely cold it is dry, and little snow falls.

(19) On a map of Europe mark and name all the places men-tioned in the tables on page 15. Beside each place state its mean January temperature. (20) Give your completed map a suitable title and write a reasoned explanation of the facts it portrays.

In summer the Eurasian land-mass *heats* rapidly, and air temperatures rise accordingly. Counting March–April–May as spring and September–October–November as autumn, (21) use the tables (*top opposite*) to write notes comparing the rate of *change* in mean temperatures in Valentia and Kursk in (a) spring and (b) autumn. Account for any differences noted.

The hot summers of Eastern and Central Europe cause convec-tional rainfall in that season, but generally speaking the total mean annual rainfall diminishes the farther east one travels into Europe. ((22) *Why?*) Indeed, the region bordering the north of the Caspian Sea is semi-desert.

Influences of depressions ('*lows*'). Cyclonic **depressions** (IV: 56) approach Europe from the Atlantic Ocean throughout the year. They contain two main streams of air: one from the south-west or south, bringing warmth and usually rain; the other from the north-west, north or east, bringing cold, blustery and showery weather. Depressions usually cover an area of several thousand square kilo-

EUROPE: Principal Depression Tracks

JAN.

COLD CONTINENT
HIGH PRESSURE AREA

JULY

S. Russia

HOT CONTINENT
LOW PRESSURE AREA

Temp. °C.	J	F	M	A	M	J	J	A	S	O	N	D
Valentia	8	7	8	9	12	14	16	16	14	12	9	8
Kursk	−9	−8	−3	5	13	17	19	18	12	6	−2	−7

metres and bring rapidly changeable, wet weather to places over which they travel. The maps below show the tracks followed by the centres of depressions crossing Europe. Study them carefully and then (*23*) write the following correctly:—

The Westerlies, in common with all other planetary wind-belts, shift $\frac{southwards}{northwards}$ in July and $\frac{southwards}{northwards}$ in $\frac{July}{January}$. Consequently in $\frac{summer}{winter}$ most depressions pass over North-west and Northern Europe, although the $\frac{high}{low}$ pressure conditions over the $\frac{hot}{cool}$ Continent in that season *do* allow occasional rain-bearing 'lows' to penetrate as far inland as Southern Russia. In winter, by contrast, changeable depression weather affects $\frac{the\ whole}{most}$ of Europe. Now, however, very cold, heavy air lying over the $\frac{west}{east}$ of the Continent hinders the eastward movement of depressions and some are deflected southwards into the $\frac{Mediterranean}{Caspian}$ Sea.

Influences of the belt of sub-tropical calms and droughts. To the south of the Westerlies lies the **planetary high-pressure belt** of sub-tropical calms (*V:42*). Here the air is constantly sinking towards the ground and warming as it does so. Hence it is relatively very dry and skies are usually cloudless. In places like the Sahara, where sub-tropical calms lie over the land throughout the year, desert conditions result. As with the Westerlies, however, the belt swings slowly northwards and southwards with the seasons. In summer it regularly encroaches over the Mediterranean and occasionally affects much of Western and Central Europe. In Mediterranean lands, therefore, summers are hot, dry and brilliantly fine, with blue skies vividly reflected in a clear sea. In winter, however, the belt of calms swings southwards and is replaced by wet, westerly winds bringing depressions. Mediterranean countries are therefore characterised by hot, dry summers and warm, wet winters; in places the rainfall maximum is in spring or autumn.

(*24*) Draw climate charts for Palermo and Cairo and account for (*a*) the hotter summer and (*b*) the lower mean annual rainfall of the latter city. (*Hints: latitude; prevailing winds.*)

PALERMO	J	F	M	A	M	J	J	A	S	O	N	D	Year
Temp. °C	11	11	13	15	18	22	25	25	23	20	15	12	—
Rain mm.	98	83	70	65	33	15	8	15	38	98	98	113	734
CAIRO	J	F	M	A	M	J	J	A	S	O	N	D	—
Temp. °C	11	13	16	20	23	26	27	27	25	22	18	13	—
Rain mm.	10	5	5	3	0	0	0	0	0	3	3	5	34

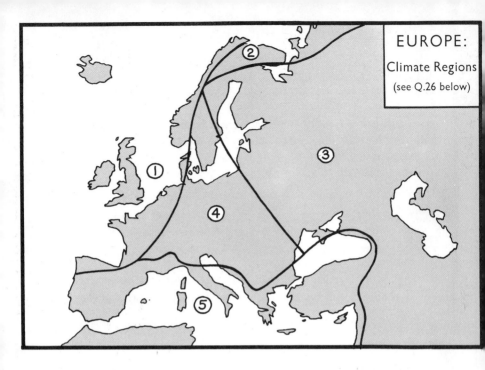

Climate Regions of Europe. This map has been divided to show the five climate regions of Europe, within each of which the seasonal pattern is more or less uniform. Referring to pages 14 to 17 and to the climate statistics on pages 72, 77 and 87 (*25*) sort out the following 'heads and tails' and (*26*) use your completed answers to label appropriately each climate region on a *large* copy of the map.

North-west European 'British' Climate		Prolonged, severe, 'Polar' winters. Warm, brief summers. Little rain, most falling in summer. Too cold for trees.
Mediterranean Climate		Mild winters. Warm summers. Ample rain all the year.
Central European 'Transitional' Climate	?	Cold winters, warm to hot summers. Moderate rainfall, most falling in summer.
East European Climate		Hot summers with drought. Warm, wet winters.
Tundra		Summers warmer and winters colder than those of Paris, but temperature extremes less than those of Moscow.

18

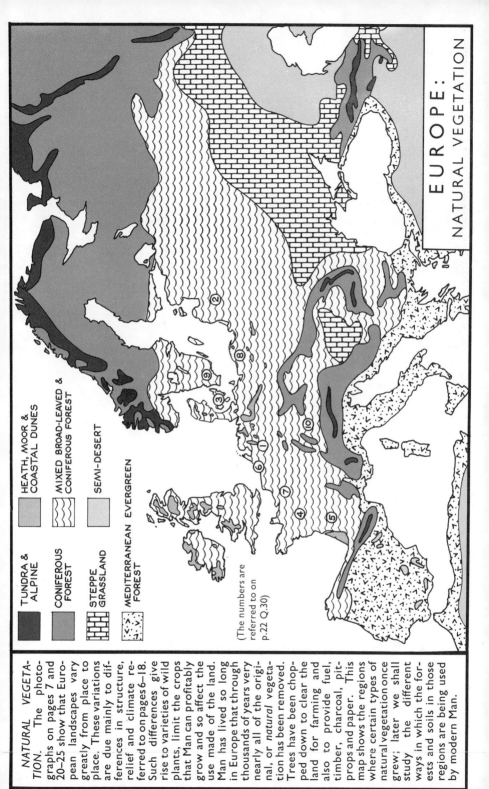

EUROPE: NATURAL VEGETATION

NATURAL VEGETATION. The photographs on pages 7 and 20–25 show that European landscapes vary greatly from place to place. These variations are due mainly to differences in structure, relief and climate referred to on pages 6–18. Such differences give rise to varieties of wild plants, limit the crops that Man can profitably grow and so affect the use made of the land. Man has lived so long in Europe that through thousands of years very nearly all of the original, or *natural* vegetation has been removed. Trees have been chopped down to clear the land for farming and also to provide fuel, timber, charcoal, pit-props and paper. This map shows the regions where certain types of natural vegetation once grew; later we shall study the different ways in which the forests and soils in those regions are being used by modern Man.

(The numbers are referred to on p.22 Q.30)

TUNDRA & ALPINE

CONIFEROUS FOREST

STEPPE GRASSLAND

MEDITERRANEAN EVERGREEN FOREST

HEATH, MOOR & COASTAL DUNES

MIXED BROAD-LEAVED & CONIFEROUS FOREST

SEMI-DESERT

A

The **tundra** of Europe is a region of 'cold desert' bordering the White Sea and Arctic Ocean. Your atlas and the map on page 19 show the countries involved; (27) name them. The vegetation here is adapted to bitterly cold, strong winds, long alternating periods of darkness and light, and a permanently frozen sub-soil. The growing season lasts less than two months, but in that time plants grow almost continuously in the long Arctic days. Typical tundra vegetation consists of dwarf, shallow-rooted shrubs, some stunted birch or willow trees and a carpet of mosses, lichens and herbaceous plants. The latter are ablaze with brightly coloured flowers in the growing season. (11:21 and 24.)

Cultivation is impossible, and the only farming use made of these barren, dreary wastes is as summer pasture for reindeer, tended by nomadic herdsmen. **A** shows a typical summer scene in Arctic Finland. This Lapp, standing beside a temporary shelter built of skins and brushwood, is well wrapped against the cold.

Despite the grim climate over 597 000 people live in the Soviet ports of Archangel and Murmansk (p. 224). Read page 14 again and (28) explain why Murmansk is ice-free throughout the year.

A 'tongue' of tundra protrudes along the Scandinavian Plateau almost to the coast of South-west Norway. (29) Why? Conditions somewhat similar to tundra are also found just below the permanent snow-line on the high peaks of Central and Southern Europe. The photograph on page 133 shows Alpine pasture at Lauterbrunnen in the Swiss Alps.

B shows part of the great **Coniferous forest** (taiga) of Northern Scandinavia, Finland and North European Russia, the only parts of Europe still extensively covered by trees. These consist mostly of pines, spruces, larch and firs, and are built to withstand harsh climatic conditions—bitterly cold, long winters; brief, cool summers; a light summer rainfall and winter snow. North of about 60° the conifers stretch away in an almost uninterrupted belt until they gradually become stunted and peter out in the Arctic wastes of the tundra. The growing season is short, but being mostly evergreens the trees are able to start growing as soon as temperatures rise above 6°C. In winter, when the ground is frozen and no food can be obtained from the soil, the trees live on food stored up in their wood. **Transpiration** in the strong winds is reduced by the hard, needle-shaped leaves. The conical shape of the trees gives them stability and prevents too

Coniferous forest near Rjukan, Norway.

B

much snow accumulating on the branches. On the forest floor a thick layer of pine-needles, together with the long, hard frosts, keeps undergrowth thin and patchy (II: 108–10).

Over wide parts of the northern coniferous forests large numbers of trees of the same species grow close together in **stands**, forming Europe's chief timber reserves. Forest products are very important to Finland, Sweden and the U.S.S.R. (pp. 42, 47 and 224).

In some places, notably along the Baltic coastlands, patches of forest have long ago been cleared for agriculture. Many farmers here combine forestry with the cultivation of hardy crops such as barley, oats and potatoes. Elsewhere in Europe coniferous trees are found in two main locations:—(i) on higher parts of the mountains of Central and Southern Europe and (ii) on poor, sandy, lowland soils where little else will grow.

Harvesting barley near Chartres in the Paris Basin.

Before they were cleared for agriculture most European lowlands were almost completely hidden by **Mixed Deciduous Forest** (map, p. 19). *Today only small patches of forest remain. Photograph* **C** *shows part of the land is now covered by intensively cultivated fields, but the surviving woodlands give the clue to its former appearance. The great mixed forest contained mostly deciduous trees such as oaks, poplar, elm, sycamore, beech and maple, although some conifers grew on poorer soils and hill slopes. The moist summer with a growing season of at least five months allows annual leaf growth, but during winter deciduous trees adapt to the cold by shedding their leaves and lying dormant.* Important agricultural regions formerly covered by deciduous forest are numbered 1–10 on the map on page 19. (30) *With the help of your atlas identify these regions from the following list:—*

Aquitaine Basin, N. Belgium and Holland, S. Germany, North-Rhine Westphalia, Poland, N.W. Russia, Paris Basin, Denmark, S. Sweden, Loire Valley

The cleared land is farmed in various ways, according to local relief, soil and climate. Major farming activities include: dairying, beef-cattle rearing and cultivation of cereals, sugar-beet, potatoes and deciduous fruits (e.g. apples, pears, plums). Intensive market gardening is found near all large industrial districts within this region.

Harvesting wheat on a collective farm in the Ukrainian Steppes.

Fifty years ago all land shown in **D** was covered with wild grasses and herbaceous plants, the natural **Steppe Vegetation** of the gently rolling plains of Central Hungary, the Black Sea coastlines and the Ukraine. South and east from the deciduous forest region the total mean annual rainfall steadily decreases until, below 500 mm, it becomes too dry for trees. Grass, however, can avoid the effect of drought by producing seeds and withering. Thus here, before Man cleared the land, stiff blades of bluish-green grasses shot up during the spring rains, but soon changed to yellow 'straw', tipped with feathery spikes, in the dry, burning heat of late summer. The grass then seeded and the seeds remained dormant during the cold winters. Year after year the old grass decayed, adding layer upon layer of rich **humus** to the soil, so the latter became black and extraordinarily fertile. The Russian name for this soil is **chernozem**.

In the late 19th century the population of Europe was increasing so quickly that new sources of food became essential. Railways stretching out across the Steppes made it possible to farm the fertile chernozems at a profit, because now for the first time the crops could be carried away cheaply to Russian cities and to the Black Sea ports for export. Today very little natural steppe vegetation remains. The wetter western districts form part of the 'maize and wheat belt' of Europe, looking very like the Prairies of the American mid-west. In the drier east and south irrigation helps the cultivation of sunflowers, barley and drought-resistant varieties of wheat, and huge numbers of cattle and sheep are reared.

Terraced hillside in Liguria, N. Italy.   Vines and vegetables are being grown.   Notice the sub-tropical trees giving shade around the house.

Mediterranean Vegetation consists of broad-leaved evergreen trees, shrubs and flowering plants, but much of the original cover of evergreen oak woodlands has long since been removed, mostly for fuel and charcoal. Its place has been taken by a **secondary growth**, known as maquis or macchia, consisting of shrubs such as oleander, rosemary, lavender and vines. Trees which survive include Corsican and Aleppo pines, chestnut and the wild olive. All these plants are specially adapted to a summer drought. In some, transpiration is reduced by small, hard leaves, thick rough bark and compact, woody stems. In others the roots are long and penetrating, or fleshy and bulbous and able to store water. Slow growth continues throughout the year. The scorching summer drought makes grass uncommon in Mediterranean lowlands, but it is found on cooler, high mountain slopes, where the yearly growth is stimulated by melting snow.

A large atlas map of Italy or Greece or Algeria will show that there is much hilly land on the shores of the Mediterranean. Consequently many of the fields are rocky and tiny and have to be cultivated by hand. This hand labour, coupled with a complete dependence on irrigation during the summer drought, gives Mediterranean agriculture a distinctive pattern. Away from the narrow coastal plains much of the cultivated land is found on carefully terraced, irrigated hillsides, as shown in **E**. As mean temperatures never fall below 6°C, plant growth is possible at all seasons. Thus most peasants grow winter crops of wheat and barley, as well as peas, beans and lentils. Mediterranean countries are also famous for their vineyards, olive groves, citrus fruit and peaches: walnuts, sweet chestnuts and almonds are also widely grown. The photo on page 198 shows a typical Mediterranean farm holding. Notice the arable patch on the valley floor, the terraced slopes and the maquis and rough pasture on the hills (11 : 80–91).

F

Heathland and Moorland Vegetation.

F shows part of the great Lüneburg Heath in Northern Germany. The map on page 19 shows that this and similar heathlands lie within a part of Europe formerly covered by forest. All of them are found on sandy, gravelly soils, such as coastal sand-dunes and glacial outwash deposits. When the original tree cover was removed rainwater rapidly soaked into the sand, carrying away valuable plant nutrients in solution. Soil fertility diminished, trees found it difficult to grow and heath plants such as heather, gorse and coarse grasses gradually took over.

The Auvergne moorland scene (**G**) has a similar air of desolation. Moorland vegetation develops on badly drained uplands where rocks are hard and impervious and the rainfall heavy. It consists of plants such as bog-moss, cotton-grass and moor-grasses, the remains of which accumulate as peat in shallow, swampy depressions. Moorlands can be converted into upland pasture by drainage and liming, and sandy soils made cultivable by heavy application of fertilisers. Such methods are costly, however, and often the only profitable way to utilise these inferior soils is to plant them with conifers.

G

POWER, INDUSTRY AND PEOPLE. On the last few pages we have seen how the natural landscape of Europe was altered by Man's attempts to wrest a livelihood from the forests and the soil. To complete the land-use picture, mention must also be made of **urban land,** i.e. ground covered by cities, coal-mines, factories, roads and so on. The maps show that the location of urban land in Europe is closely related to the distribution of minerals, especially coal and iron ore. The reasons for this date back about 250 years—to the first successful smelting of iron ore using coke instead of charcoal in the blast furnace, and to the invention of the steam engine. Both these key discoveries were made in Britain, and it was there that they first began to revolutionise industry. Coalfields rapidly became major industrial centres. Factories of all kinds were built on them so as to be near the source of power to drive their machines; iron and steel mills needed coking coal; and metal-working engineering industries were located close to the supplies of iron and steel. Thus developed the grimy, noisy, bustling industrial towns of the 'Steel Age'. Into them flocked tens of thousands of people from the surrounding countryside.

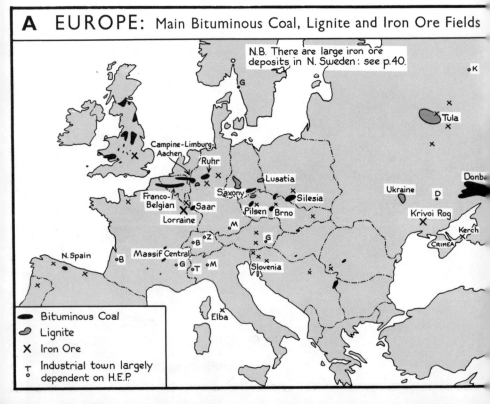

A EUROPE: Main Bituminous Coal, Lignite and Iron Ore Fields

N.B. There are large iron ore deposits in N. Sweden: see p.40.

- Bituminous Coal
- Lignite
- X Iron Ore
- Industrial town largely dependent on H.E.P.

The new ideas spread in time to the Continent, and the maps show the effects of coalfield industrial development on the distribution of population. Map **A** shows the main bituminous coal and lignite deposits in Europe. Notice again that most of them lie along the northern foothills of Hercynian mountains such as the Ardennes, Lower Rhine Highlands and Erzgebirge. ((*31*) Which coal deposits are *not* so located?) The dense numbers of people in this west–east belt mostly live in industrial towns built on or near the coalfields, but remember that this is also the most intensively cultivated and thickly peopled farmland in Europe. ((*32*) Why? See page 11 again if necessary.)

The European coalfields did not all undergo industrialisation at the same time. First was the *Franco-Belgian coalfield*, which developed rapidly from the late 18th century onwards, in some cases under the direct supervision of skilled British craftsmen and engineers who settled there. One such was John Cockerill, who in 1832 founded a famous iron and steel firm near Liège—today the Cockerill industrial 'empire' is still pre-eminent in Belgium. *German coalfields* did not become major industrial centres until late in the 19th century. The main reason was that until then the

B   EUROPE: Major Cities and Very Densely Populated Regions

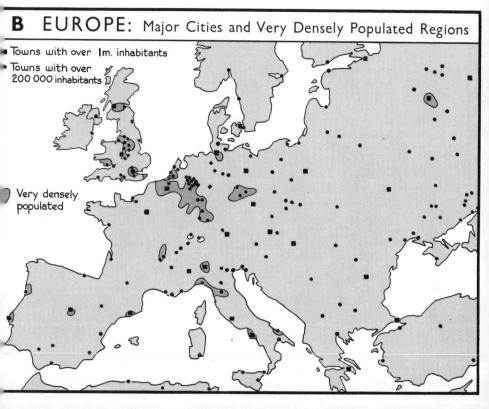

Towns with over 1m. inhabitants

Towns with over 200 000 inhabitants

Very densely populated

German lands were divided into a large number of small, independent and economically weak kingdoms. Unification under Prussia led to the creation in 1870 of a powerful nation-state (*p. 30*), and from then until the First World War Germany went all out to build up her heavy industry. ((*33*) Who was Bismarck? Find out what you can about his policy of 'Blood and Iron'.) From 1871 until 1918 Germany controlled the great Lorraine iron ore field, the ore from which helped to build up the immensely important steel and heavy engineering industry of the Ruhr coal-mining area. The latter soon became one of the world's major industrial regions. Further details of German coalfields are given on pages 96–99 and 232.

After the First World War other coalfields became centres of growing industries, especially in *Poland, Czechoslovakia* and *European Russia*. (*34*) Name the four main coalfields concerned, using the map on page 26 and your atlas. Of these the Donbas was destined to become the principal heavy industrial region in the Soviet Union, using local coal and iron ore brought from Krivoi Rog and the Crimea.

Since the late 19th century new sources of power have become

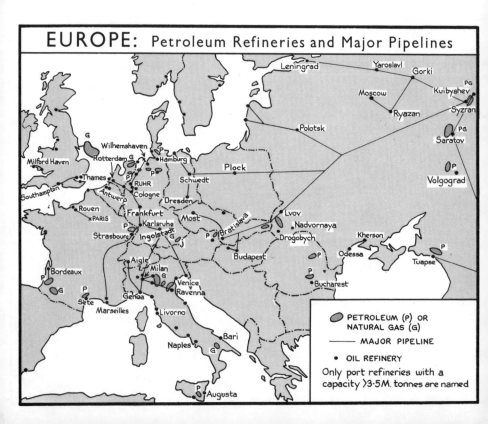

EUROPE: Petroleum Refineries and Major Pipelines

available. Electricity, mainly generated in coal-fired power stations and distributed by 'Grid' cables, allows factories to be sited more or less anywhere. In some countries, such as Switzerland, Italy, Norway and Sweden, running water is harnessed to generate hydro-electricity; specialised industries including electrical engineering, electro-chemicals and electro-smelting have grown up as a result at the source of cheap electrical power. Indeed Northern Italy has become a major industrial region. Certain industrial towns largely dependent on supplies of hydro-electricity are indicated on the map on page 26: (35) name each of these towns and state which country each is in. In the past 50 years, too, petroleum and natural gas have become ever-increasing sources of power for industry and transport. Western Europe now consumes about 20% of the entire world output of petroleum, but only a small fraction is produced within that Continent. On the other hand very large deposits of natural gas have recently been discovered in the Netherlands, Italy, the Soviet Union and under the North Sea: the map shows the main European oil- and gas-producing centres. The 500 million tonnes of petroleum imported annually into Europe come mainly from the Middle East and Venezuela, but France imports increasing quantities from North Africa. Until the 1950's most European oil refineries were built at ports: the principal ones are indicated on the map. (36) Name each of them and say which country each is in. In the past few years many very large *inland* refineries have been constructed, close to important 'markets' for oils and petrols. For example, crude oil is pumped inland from Wilhelmshaven, Rotterdam and Marseilles, chiefly to the Rhine Valley industrial cities. Pipelines also link the Soviet oilfields at Second Baku to industrial centres in Eastern Europe. (37) Name the main oil-refining towns using this petroleum and state which country each is in.

Until the end of the Second World War in 1945, most of the very large-scale industry of Europe was confined to western and north-western countries. The one important exception was the Soviet Union, where industries were developed with remarkable speed after the Communists seized control there in 1917. Since 1945, however, great changes have taken place. They stem from the fact that Hitler's Germany was destroyed by attacks from both east and west, and the final battles brought Russian troops as far west as Berlin. Communist governments, based on that of the Soviet Union, were installed in the territories of Eastern Europe

occupied by the Red Army, and Czechoslovakia
became Communist in 1948. Under Russian
influence the Communist states have made tremendous efforts to stimulate industry, all available
supplies of coal, lignite, water power and mineral
ores being marshalled through a succession of
'Five-Year Plans'. Similar State-aided attempts
to develop industry have been made in Jugo-
slavia, Spain and Southern Italy. As a result the
traditional way of life of many people in Eastern
and Southern Europe is changing, in some places
very quickly.

For centuries the great majority of them have
been poor, uneducated peasant farmers, their
whole lives devoted to eking out a frugal living
for themselves and their families from small, often
scattered plots of land. Living standards in these
largely backward agricultural countries have
lagged far behind those of industrial nations such
as Belgium and Germany. An industrial nation
is able to produce a great range of goods and services designed to make life pleasanter and less
arduous. Think of the effects of this in your
own homes: electricity, mains drainage, mass-
produced and cheap furniture, utensils, clothes,
washing machines, television sets and so on. In
addition an industrial country can export some
of its manufactures in exchange for foodstuffs,
industrial raw materials and other manufactures
from other parts of the world. The resulting
increase in national wealth eventually benefits
all the population, in the form of higher wages,
better food, more goods in the shops and improved public services. Where fuel and industrial raw materials are both lacking, attempts can
still be made to develop agriculture and so to
build up wealth by trade in farm produce. In
the course of this book we shall come across many
examples of both industrial and agricultural
development schemes which are greatly altering
the geography of parts of modern Europe.

NATION - STATES
OF EUROPE AND
THEIR GROUPINGS.
A nation-state may
be defined as a
specific territory,
inhabited by people
owning allegiance to
a single, sovereign
government. This
map shows the
twenty-four more
or less independent
states of modern
Europe. A glance
through an histori-
cal atlas will show,
however, that the
present political
pattern has existed
only since 1945, i.e.
since the end of the
Second World
War. For hun-
dreds of years be-
fore that date the
boundaries of the
various states were
constantly altering,
usually as a result of
lengthy periods of
warfare. Some states
have changed more
than others, and
some, e.g. Poland,
have at times disap-
peared as their ter-
ritory was absorbed
by more powerful
neighbours. Of the
present more
powerful European
states only Britain,
France and Russia
had grown to ap-
proximately their
present shape by the
mid-19th century.

Throughout his-
tory, European
states have formed
pacts of allegiance
to give themselves
added protection in
the event of war.
These pacts have
altered as the power
and fortunes of the
various members
have fluctuated.
Since 1945 Europe

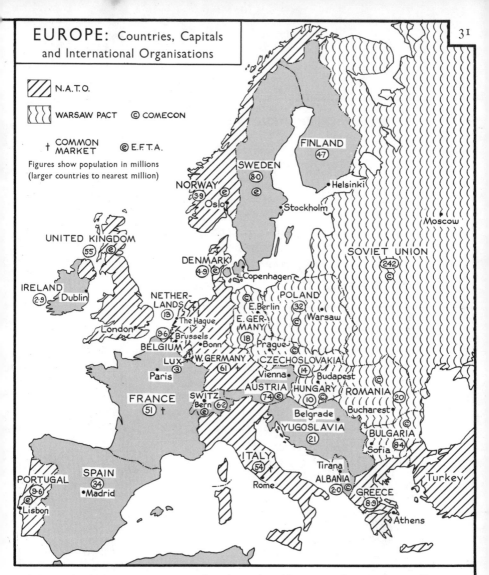

EUROPE: Countries, Capitals and International Organisations

31

N.A.T.O.

WARSAW PACT   © COMECON

† COMMON MARKET   Ⓔ E.F.T.A.

Figures show population in millions
(larger countries to nearest million)

FINLAND (4·7)
•Helsinki

SWEDEN (8·0)

NORWAY (3·9) Ⓔ
•Oslo

Stockholm•

Moscow•

SOVIET UNION (242)

UNITED KINGDOM (55) Ⓔ

DENMARK (4·9) Ⓔ
Copenhagen•

POLAND
E.Berlin• (32) ©
Warsaw ©

IRELAND (2·9) •Dublin

NETHER-LANDS (13) Ⓔ
•The Hague
•Brussels

E.GER-MANY (18)
Prague ©

London•

BELGIUM (9·6) Ⓔ
•Bonn

CZECHOSLOVAKIA (14)

LUX. (·3)
W. GERMANY (61)
Vienna• •Budapest

•Paris

AUSTRIA (7·4) Ⓔ
HUNGARY (10) ©

ROMANIA (20) ©

FRANCE (51) †

SWITZ. (6·2) Ⓔ
Bern•

Belgrade•
Bucharest•

YUGOSLAVIA (21)

BULGARIA (8·4)

ITALY (54)
Sofia•

PORTUGAL (9·6) Ⓔ
•Lisbon

SPAIN (34)
•Madrid

Rome•

Tirana•
ALBANIA (2·0) ©
GREECE (8·9)

Athens•

Turkey

has been divided into two main political groups or 'blocs', viz. the Communist countries of Eastern Europe, and the non-Communist states in the west and south.

The Communist countries, dominated by the U.S.S.R., are linked by the Warsaw Pact, a military treaty of mutual assistance signed in Warsaw in 1955. ((*38*) *Name its members.*) To counter a possible attack by the Communist countries several of the non-Communist countries formed, in 1949, an alliance with the U.S.A. and Canada called N.A.T.O. (*39*) Give the full name of this organisation and name its European members. (France, a founder member, withdrew from N.A.T.O. in 1966.)

In addition to these military alliances, various economic groupings have been formed for purposes of trade. The most important of these are:—(i) *The European Common Market* set up by ((*40*) *which?*) countries in 1959. Tariff barriers between these countries will eventually be removed, after which they will form the world's largest free trade area. (ii) *The European Free Trade Area*, formed in 1961 on the initiative of the United Kingdom. (*41*) Name its other members. (iii) *Comecon*, the Russian-dominated organisation concerned with the economic development of Eastern Europe.

CHAPTER 2

Northern Europe:
Norway, Sweden, Finland, Denmark and Iceland

THE SCANDINAVIAN PENINSULA consists of Norway and Sweden. The maps, diagrams and photos here and on pages 34–6, together with the map on page 19 and the climate figures on page 15, tell us much about its geography. Study them carefully and complete the following exercises:—

(*1*) *Write a geographical account of a journey from Bergen to Söderhamn. In your description pay particular attention to the geology, scenery and human activities (if any) of the regions through which you would pass.*

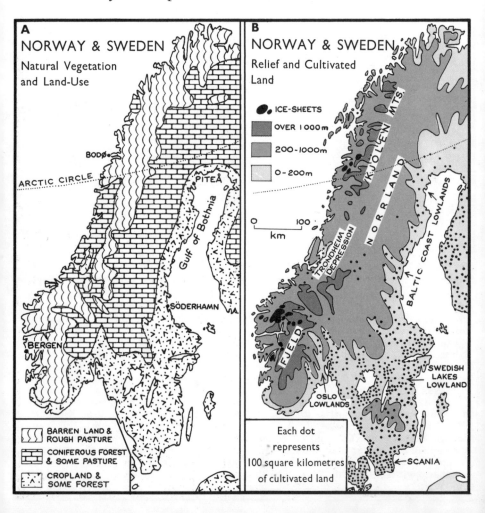

A
NORWAY & SWEDEN
Natural Vegetation and Land-Use

BODØ

ARCTIC CIRCLE

PITEÅ

Gulf of Bothnia

SÖDERHAMN

BERGEN

BARREN LAND & ROUGH PASTURE
CONIFEROUS FOREST & SOME PASTURE
CROPLAND & SOME FOREST

B
NORWAY & SWEDEN
Relief and Cultivated Land

ICE-SHEETS
OVER 1 000 m
200-1000 m
0-200 m

KJØLEN MTS.

NORRLAND

0   100
km

TRONDHEIM DEPRESSION

BALTIC COAST LOWLANDS

FJELD

SWEDISH LAKES LOWLAND

OSLO LOWLANDS

SCANIA

Each dot represents 100 square kilometres of cultivated land

(2) *Explain why the entire coast of Norway is always ice-free, whereas the Gulf of Bothnia is frozen for several months each winter.*

(3) *Why is the total mean annual rainfall at Piteå so much less than that at Bodø?*

(4) (a) *Describe the distribution of population in Norway and Sweden.*

    (b) *Give four reasons why so much of these countries is virtually uninhabited.*

(5) *Suggest reasons why (a) farming is of much greater importance in Sweden than in Norway and (b) the people of Norway depend much more on fishing.*

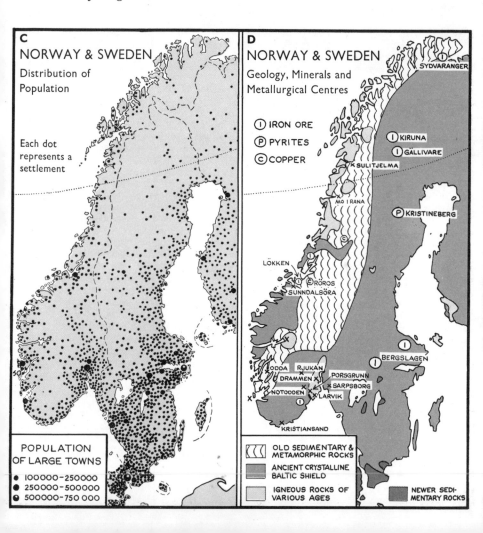

C

NORWAY & SWEDEN

Distribution of Population

Each dot represents a settlement

POPULATION OF LARGE TOWNS
- 100000–250000
- 250000–500000
- 500000–750000

D

NORWAY & SWEDEN

Geology, Minerals and Metallurgical Centres

- Ⓘ IRON ORE
- Ⓟ PYRITES
- Ⓒ COPPER

SYDVARANGER
Ⓘ KIRUNA
Ⓘ GÄLLIVARE
✕ SULITJELMA
MO I RANA
Ⓟ KRISTINEBERG
LÖKKEN
Ⓒ RÖROS
SUNNDALSÖRA
BERGSLAGEN
ODDA
RJUKAN
DRAMMEN
PORSGRUNN
NOTODDEN
✕ SARPSBORG
Ⓘ LARVIK
KRISTIANSAND

OLD SEDIMENTARY & METAMORPHIC ROCKS
ANCIENT CRYSTALLINE BALTIC SHIELD
IGNEOUS ROCKS OF VARIOUS AGES
NEWER SEDIMENTARY ROCKS

A

A shows part of the rugged **fjord coast** of Norway. Indentations such as this are found along the entire coast from Stavanger to the North Cape. These gigantic furrows are U-shaped glaciated valleys, gouged during the Ice Ages by ice spilling westwards over the steep edge of the Scandinavian Plateau. The valleys were later partially submerged by the post-glacial rise in sea-level. Fjords are remarkable for their steep, often precipitous sides as well as for their great length and narrowness. One of the longest is the Sogne Fjord, which although 179 km long is rarely more than 5 km across. The water in the Sogne Fjord is in places over 1300 m deep, and the fjord walls rise abruptly a further 1600 m to the plateau summit. Near the entrance to each fjord is a bar or **threshold** where the water shallows to about 50–200 m.

Running parallel with the mainland is a string of some 150,000 islands, most of which are low, rounded hummocks called **skerries**. The islands form a natural breakwater called the **Skerry Guard**, behind which is a calm channel forming a magnificent north–south

routeway for coastal steamers. Settlements along the fjord coast are few and far between. Most people live in fishing ports such as Bergen (p. 38) and the rest in isolated farm villages built on alluvial flats at the fjord heads, or on small terraces of level ground near their mouths.

The central 'spine' of Scandinavia consists of the higher ridges of a plateau, gently tilted towards the east and much dissected by river and glacial erosion. Dissection is in fact so far advanced that in places the high plateau has been reduced to a series of isolated, rounded summits, all at approximately the same level. (6) Explain this process by means of a labelled diagram (V:31). North of Trondheim the plateaus form a long, narrow ridge. (7) Name it (map, p. 32).

To the south the mountains are higher, and their broad summits are known as the **fjeld**. Here is one of the most desolate regions in Europe, a barren, boulder-strewn, uninhabited wilderness still covered in parts by extensive snow-fields, small ice-sheets and glaciers. (8) From the map on page 32 estimate the

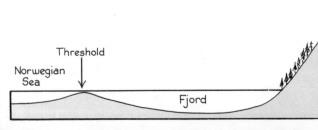

Glaciated Fjeld

Nunatak

Threshold

Norwegian Sea

Fjord

Nunataks are peaks which protruded through the ice-sheets. They were affected by frost, but not by glacial erosion. Thus in contrast to the rounded appearance of most of the fjeld the Nunataks are jagged and angular.

area of the largest of these ice masses. **B** shows part of the ice-sheet near Finse, on the Hardanger Fjeld. (9) Make an outline sketch of this photo, and in the places marked A–F add appropriate labels from the following list:— ice-sheet/level plateau summit/**terminal moraine**/melt-water lake/**ground moraine**/crevasses.

Occasional bold peaks such as the 2699 m Galdhöppingen stand out above the gently undulating surface of the fjeld. These peaks were probably **nunataks** protuding through the Quaternary Ice-sheets (see diagram below).

On the high fjeld the only vegetation consists of patches of tundra, but poor pasture grows on the lower slopes of the mountains and trees appear as the level of the plateau gradually drops below 700 m. Towards the east the trees thicken into dense coniferous forests: (Photo and description, p. 20). They grow on infertile soils covering sheets of glacial gravels, sands and clays: occasional

gaps in the forest show where the underlying crystalline rocks were swept bare by ice erosion.

The main occupations on the dip slope of the plateau are lumbering and mining. Forestry is assisted by the large number of approximately parallel rivers draining eastwards to the Baltic. These rivers provide hydro-electricity for saw and pulp mills, and logs can be floated downstream after the spring thaw.

Many isolated mining camps dot both the taiga and the tundra to the north, for the Baltic Shield (p. 6) is rich in both ferrous and non-ferrous minerals. The location of the main workings are shown on map D, page 33. The most important deposits are the iron ores at Gällivare and Kiruna in Northern Sweden, and in the Bergslagen district to the north of the lakes of Central Sweden. Huge quantities of ferrous and non-ferrous ores are sent to Baltic ports such as Luleå, Gälve and Oxelösund for refining and export.

Bare Rock & Rough Pasture

Dense Coniferous Forests
Some Farm Clearings

H.E.P. Station and Chemicals Factory at Rjukan, Norway. (See also map and p. 39.)

THE BALTIC COAST PLAIN is one of the few places in Scandinavia really favourable to agriculture. Deposits of geologically recent silts here cover the ancient rocks of the Shield, and soils are relatively fertile. They are cultivated despite the long hard winter and short growing season. Mixed farming predominates: barley and rye are the main cereal crops and roots, potatoes and hay are grown as animal fodder.

Pastoral farms are also found in valleys such as the Glåma and Lågen which trench the southern and eastern edges of the plateau. In summer, when the snow has melted, cattle and sheep from many valley farms are driven up to graze on the poor pastures of the lower fjeld. Here they remain in temporary summer holdings or **saeters** for about 10 weeks, returning home before the onset of the autumn snows. Some saeters are as far as 112 km from the valley farms which use them. **Transhumance** such as this emphasises the desperate lack of farmland, especially in Norway.

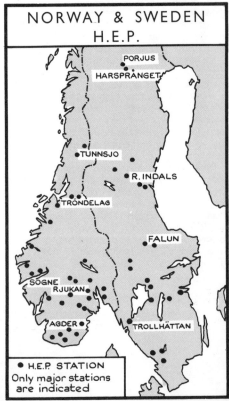

NORWAY & SWEDEN H.E.P.

PORJUS
HARSPRÅNGET
TUNNSJO
R. INDALS
TRONDELAG
FALUN
SÖGNE
RJUKAN
AGDER
TROLLHÄTTAN

● H.E.P. STATION
Only major stations
are indicated

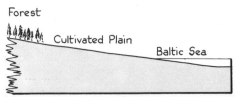

Forest
Cultivated Plain
Baltic Sea

Cultivable Forested	Barren (Fjeld, Ice-cap & Tundra)

NORWAY covers an area about the same as that of the British Isles, but whereas the latter's population is 58·2 million, only 3·9 million people live in Norway. The main reasons why Norway's population is so small and unevenly distributed (map **C**, p. 33) are the barrenness and lack of resources of most of the country, i.e. the land has never been able to *support* a large population. (*10*) From the diagram above estimate the proportion of Norway covered by each of the following: *barren fjeld; forested land; cultivable land.*

Farming. One in three Norwegians relies entirely or mainly on farming for a livelihood. Another tenth of the population ((*11*) *i.e. how many people?*) live on farms, but treat farming as a subsidiary occupation and spend most of the year in mining, forestry or fishing. The main farming districts are the lowlands around Oslo, Trondheim and the Jaeren Plain south of Stavanger. As well as having the only large patches of fertile soils, these lowlands have the longest growing season in Norway and easy access to large urban markets in and around Oslo. Isolated patches of farmland are also found as far north as Finnmark, but towards the north the growing season gets progressively shorter and preparing land for tillage becomes a frantic rush against time. Particularly suitable for cultivation are the less steep, south-facing slopes of many fjord walls. (*12*) Why? (*Hints: shelter, sun.*)

In general the wet, equable climate and scarcity of fertile lowland favour pastoral rather than arable farming. Hence nearly all Norwegian farmers obtain a cash income by selling milk. Hay, oats, barley and roots are grown as animal feed, and large quantities of potatoes for human consumption. (*13*) Suggest reasons why (*a*) farmers in Western Norway have difficulty in drying their hay crop, whereas in Eastern Norway most pastures are irrigated, and (*b*) the only really important grain-growing area lies around Lake Mjøsa.

The very scanty amount of fertile land has always affected the way of life of Norwegians, for many of them have been forced to seek a livelihood in directions other than farming and forestry. For centuries the chief additional source of wealth was the sea. More recently the development of Norway's enormous hydro-electric power resources made possible the growth of manufacturing industry (*details, p. 39*). The relative importance of Norway's main productive activities are indicated in the bar diagram below.

NORWAY: GROSS NATIONAL PRODUCT BY INDUSTRY (Total £4510 m.)

FISHING	FORESTRY	AGRICULTURE	BUILDING & CONSTRUCTION	TRANSPORT EQUIPMENT [including ships]	COMMERCE	MINING & MANUFACTURING	OTHER INDUSTRIES

NORWAY'S FISHING INDUSTRY

Northern Seas. Fishing mainly for COD, around Lofoten Isles and off north coast. Cod approach these waters annually from Atlantic Ocean to spawn. Main cod fishing season from January–April. Some 25 000 fishermen and 4000 boats involved, many coming from farther south. Norwegian trawlers also fish off Iceland, Newfoundland and in White Sea. Chief cod fishing ports:—Tromsö and Hammerfest. Much cod formerly dried, salted and smoked for export: now mostly deep frozen at Findus works in Hammerfest. By - products:—cod-liver oil; fish meal, fertilisers and glue from fish waste.

Southern Seas off Trondheim, Bergen and Stavanger. Fishing mainly for HERRING. More herring landed in Norway than any other fish. Largest catches between January and March. Main herring ports: Ålesund, Haugesund, Bergen. Some herring caught in all months along entire Norwegian coast. Most of catch converted into herring oil and meal. Stavanger has BRISLING fisheries. Other fish landed in Norway include HADDOCK, SPRATS and MACKEREL. Much DOGFISH (rock salmon) landed at Måløy and sent to Britain for fried-fish trade.

Average annual landings in Norway's main ports are indicated in thousands of tonnes by the figures in brackets. (*14*) Draw bar diagrams to illustrate these figures. (*15*) Draw bar diagrams to illustrate the table below. (*16*) Use the notes and figures given here to write an account of Norway's fishing industry.

Map labels:
- Hammerfest (60)
- Vardö (200)
- Tromsö (100)
- LOFOTEN ISLANDS
- Bodö (10)
- Namsos (20)
- Molde (300)
- Ålesund (300)
- Måløy (14)
- Florö (20)
- Bergen (300)
- Haugesund (20)
- Stavanger (20)
- Tönsberg
- Kristiansand (10)
- NORWAY

Whaling. Norwegians pioneered whaling in Arctic Ocean. Wholesale slaughter of whales there led to shift of most whaling to Antarctic. Here Norway operates world's largest whaling fleet. Large 'factory' vessels sail from Tönsberg, Sandefjord and Larvick in Østfold. After capture by explosive harpoons, whales are processed on board. Chief products: whale oil for manufacture of lubricants, burning oils, soap and margarine. Bones and other refuse converted into bone meal and fertilisers.

Forestry. (*17*) From the diagrams on page 37 state (*a*) the percentage of Norway's land surface covered by forests and (*b*) the percentage of Norway's national income derived from forestry and timber industries. The latter figure, coupled with the fact that these industries employ 25% of the workmen in Norway, shows their importance to the country's prosperity. The main commercial forests are the pine and spruce stands in the valleys converging on Oslo Fjord and Trondheim. In addition most

LEADING EUROPEAN FISHING COUNTRIES: TOTAL ANNUAL LANDINGS
(Thousands of tonnes)

European Russia*	2 010	U.K.	1 010	Portugal	579
Norway	1 950	Denmark	998		
Spain	1 270	France	774		
Iceland	1 086	W. Germany	628		

* Includes Black and Caspian Sea fisheries.

Norwegian farms include some forested land, and forestry adds something to most farmer's incomes. The main saw-mills, pulp- and paper-mills and export centres for wood. products are Trondheim and Skaggerak ports such as Kristiansand, Larvik and Drammen. (*18*) Why are the latter ports well suited for these functions? (*Hint: transport.*) Shipments of wood pulp, paper and board products represent 25% of the value of Norway's export trade.

Manufacturing. Norway has few minerals and very little coal, but enormous supplies of cheap hydro-electricity have favoured the development of certain modern industries, notably electro-smelting and electro-chemicals. Details are given in the table below. (*19*) Show this information on a map. (*Use map* **D**, *p. 33 and map p. 36 for location of major power plants.*) Notice on your finished map that most installations are located (*a*) close to sources of hydro-electricity and (*b*) on tide-water, to facilitate import of raw materials and export of finished products. Most mineral ores smelted in Norway are imported, but substantial deposits of iron and copper are mined at Sydvaranger and Lökken respectively. Norway is now the fourth largest producer of aluminium.

Although three-quarters of Norway's water-power resources remain unharnessed, electricity is greatly used in homes, mills and factories. Fortunately many rivers, falls and lakes suitable for future power schemes are in the south and south-east of the country where most industrial firms are already located.

Norway's traditional 'free trade' policy and small population have in the past hindered the growth of manufacturing ((*20*) *How?*) but now its membership of the European Free Trade Area (*p. 31*) has enlarged possible markets, and a big expansion is forecast for Norway's engineering industries. These lie mainly in the Oslo district and produce forestry and wood-working equipment, power plant and cables and equipment for the fishing industry. Marine engineering and ship-building are important in Stavanger, Oslo, Bergen and Trondheim. Despite its relatively small population Norway has the world's third largest merchant navy, engaged in the carrying trade all over the world. Money paid to shipping firms by foreign merchants is a leading source of revenue, for Norway's prosperity is vitally dependent on foreign trade.

NORWAY: Main electro-smelting and electro-chemical installations	
Type	*Site*
Iron and Steel	Mo i Rana
Copper	Röros
	Sutlitjelma
Ferro-alloys	Kristiansand
	Larvick
	Drammen
	Notodden
Aluminium	Sogne Fjord
	Sunnsdalsöra
Carbide	Sarpsborg
	Odda
Fertilisers	Notodden
	Rjukan
	Porsgrunn

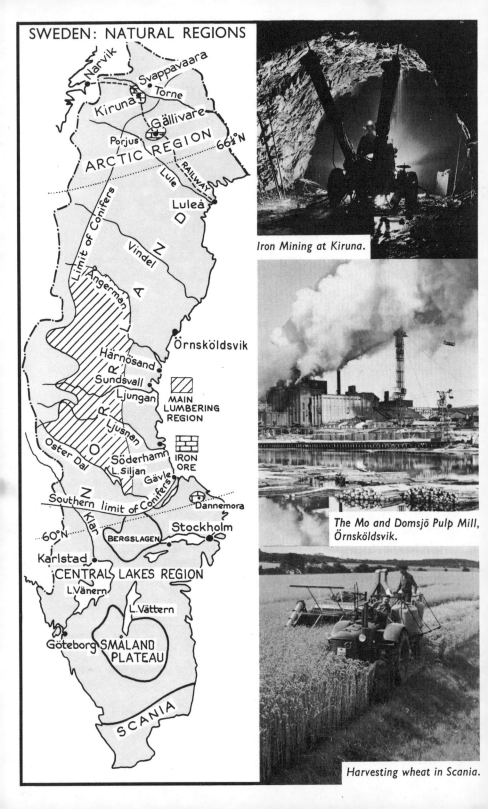

SWEDEN: NATURAL REGIONS

Narvik
Svappavaara
Torne
Kiruna
Gällivare
Porjus
ARCTIC REGION
66½°N
RAILWAY
Lule
Luleå
Vindel
Ångerman
Limit of Conifers
Örnsköldsvik
Härnösand
Sundsvall
Ljungan
MAIN LUMBERING REGION
Ljusnan
IRON ORE
Öster Dal
Söderhamn
L. Siljan
Gävle
Southern limit of Conifers
Dannemora
60°N
Klar
Stockholm
BERGSLAGEN
Karlstad
CENTRAL LAKES REGION
L. Vänern
L. Vättern
Göteborg SMÅLAND PLATEAU
SCANIA

Iron Mining at Kiruna.

The Mo and Domsjö Pulp Mill, Örnsköldsvik.

Harvesting wheat in Scania.

SWEDEN is a land of many contrasts: some parts are as intensively cultivated as any in Europe, while others are covered by dense forest or by tundra. The varied rocks, soils and landscapes are reflected in four of the main occupations of the working population, of whom 11% are employed in farming, 1% in mining, 7% in forestry and associated industries and 22% in engineering. To a marked extent each of these activities predominates in a different part of Sweden. There are five main regions (*see map*), only one of which (the Småland Plateau) is virtually useless to Man. The four productive regions are illustrated in the photographs opposite and on page 44.

ARCTIC SWEDEN can be reached by swift electric train from Stockholm in twenty-two hours.

". . . As you approach the Arctic Circle the trees become smaller and soon nothing is left but the stunted birches that cover thousands of square miles of this northern countryside. . . . Then, quite suddenly, 100 miles inside the Arctic Circle, the train runs into a large modern station complete with first-class hotel. On the hill behind stands Kiruna, a town of some 20 000 inhabitants. . . . From this point to the north nothing is to be seen but flat swampy land covered with small bushes and scrub. The view westwards is somewhat similar except that the horizon is dominated by a range of snow-covered mountains. It is when one looks south that the reason for the existence of such a large community in this barren and inhospitable land becomes apparent. Below the station lies a long lake and on its further shore another hill rises to a considerable height. It is split almost in half by a great gash many hundreds of feet deep and this gash is the world-famous iron-ore mine. Every person in Kiruna depends directly or indirectly on this great man-made hole for his or her living. Work goes on there day and night throughout the whole year, through the six weeks of perpetual sunshine in the summer and, more remarkably, through the long Arctic night and winter with temperatures down to −40° Centigrade."*

Iron-bearing rocks cover nearly 8000 square kilometres around Kiruna and the other main mining centre at Gällivare. Most of the ore is very high grade, yielding between 50%–70% iron, but a new low-grade pit at Svappavaara, yielding 35% iron, is now the third largest mine in Sweden. Exploitation is entirely dependent on the railway, which first reached Kiruna from the Baltic port of Luleå in 1892. As Luleå is ice-bound for five months another rail-link, opened in 1902, was driven across the barren Kjölen Mountains to the ice-free port of Narvik (*see map*). Since then Narvik has despatched two-thirds of the iron ore exports from

* Noel Watts, 'Kiruna: Sweden's Northernmost Mining Town', *Geographical Magazine*.

Arctic Sweden, mostly to Germany. Power for the Narvik railway comes from the great hydro-electric plant at Porjus (*map, p. 36*), and this plant also sends electricity to the town and mine at Kiruna. In mid-winter, when the sun is below the horizon for five weeks, mining continues under the glare of arc-lamps (*III:* 193).

THE NORRLAND FOREST REGION was described on pages 20 and 35. (*21*) Use that information, the map on page 40 and the figures opposite to write the following passage correctly:—

In Sweden, dense forests stretch from latitude ...°N $\frac{\text{to beyond}}{\text{almost to}}$ the Arctic Circle. Their trees are a major source of wealth, for timber and timber products make up no less than ...% of Sweden's exports Most of the trees are $\frac{\text{coniferous}}{\text{deciduous}}$, spruce being commercially the most important species. Other softwood conifers include ...., ... and .... Towards the south $\frac{\text{deciduous}}{\text{coniferous}}$ trees such as beech are felled, whilst in the extreme north the hardy b.... predominate. Throughout the forest region the timber is very accessible, due to the large number of rivers and lakes: these provide natural waterways down which logs are floated to the $\frac{\text{Kattegat}}{\text{Baltic}}$ coast and to Lake $\frac{\text{Vättern}}{\text{Vänern}}$. Rivers such as the O.... Dal and L...... also provide abundant hydro-electric power for saw- and pulp-mills. The main lumbering districts lie between the A...... River and Lake S.... Sundsvall, at the mouth of the River ...... is the main centre of saw- and pulp-mills, more than fifty such factories being located there. Other important river-mouth centres on the Baltic Coast include G...., S........ and H........ On Lake Vänern the main centre is K....... In the past $\frac{\text{saw-mill products}}{\text{wood pulp and paper}}$ were more important than $\frac{\text{saw-mill products}}{\text{wood pulp and paper}}$, but nowadays the situation is reversed; a sign of growing industrialisation in Scandinavia.

Until recently timber felling in Sweden has been almost entirely

MAIN USES OF SWEDISH TIMBER

Building materials—e.g. flooring, ply-wood, wallboard and pre-fabricated doors, window frames and houses. Exports of pre-fabricated goods much expanded in recent years.

Wood-pulp—obtained mainly by chemical treatment of spruce chips in huge vats. Final product looks like greyish-white cardboard. Exported in compressed sheets for manufacture into paper. By-products of wood-pulp production include paints, varnishes, cosmetics and ethyl-alcohol. The alcohol is blended with imported petroleum to make motor fuel.

Paper—most paper-mills located in Central Sweden where industry is more advanced (see opposite) and where the main home demand for paper lies. Emphasis on newsprint production.

Fuel and charcoal—Sweden lacks coal. Much deforestation in former times to obtain charcoal for iron and steel production. High-grade Swedish steels still made by charcoal smelting.

Matches—Sweden a leading producer. Main centre: Jönköping, where the safety match was invented.

Railway sleepers, pit-props and *telegraph poles*—from pines.

Year	1920	1935	1950	1955	1960	1962	1972
Saw Logs	460	460	540	680	760	685	440
Pulp Wood	210	460	455	770	905	1 050	1 380

a winter occupation. At that time it was easiest to recruit temporary labour ((22) *Why?*) and the icy ground surface facilitated the hauling of logs to the edge of frozen waterways. Now, with demand for wood-pulp constantly rising, large numbers of trees are felled throughout the year by a permanent labour force using portable power saws. Log transportation is changing, too, as new forest roads provide cheaper routes to the coast than the traditional floatways.

THE CENTRAL LAKES REGION is mainly low-lying, undulating and fertile, but some forested ridges of hard, ancient rocks rise above the plains. This is the most prosperous region in Sweden, with mixed farming, mining and many industries. The latter give employment to two-thirds of Sweden's 3·2M working population, and of these some 10% are in Stockholm alone. The other main industrial towns are Eskilstuna, Jönköping and Göteborg, but many smaller centres are scattered throughout the region (*see map overleaf*). The main reason for their widespread distribution is Sweden's lack of coal; no closely packed industrial 'black country' has developed. Many small industrial towns are still located at sources of water-power, the water-wheels of former times being replaced today by hydro-electricity. There are also many scattered mineral deposits in central Sweden which for centuries have been centres of mining and manufacturing (*see map*). Like Norway, Sweden has a relatively small home market, and so Swedish industrialists have always sought to sell their goods overseas. Modern Sweden's world-wide reputation for high-quality engineering products is based on technical skill.

THE SMÅLAND PLATEAU is an upland region, where ancient Shield rocks appear at the surface. Covered in places by sterile glacial sands, it rises to about 400 m like an island of infertility amidst rich farming land. The unattractiveness of this region is indicated by its very scanty population (*map C, p. 33*).

SCANIA is the most important farming district in Sweden. The underlying rock there is chalk, but a covering of boulder-clay gives rise to fertile soils. These, coupled with the relatively mild climate, favour the cultivation of wheat and sugar-beet in addition to hardier crops such as rye, oats and potatoes. Scania is

SWEDISH EXPORTS: TOTAL VALUE OF EXPORTS ABOUT £2400 m.

Machinery, Ships, Autos etc.	Iron ore, Steel etc	Paper & Pulp	Wooden Products	Miscellaneous

MAIN INDUSTRIES IN CENTRAL SWEDEN

METALLURGY AND ENGINEERING are the most important. Mineral ores, notably of copper, silver, zinc and iron, have been worked for centuries in the hilly Bergslagen district. The famous copper ores at Falun are now exhausted, but pyrites is still dug there. Today the main workable deposits are of iron ore; black magnetite at Grängesberg and red haematite (70% iron) at Dannemora. The magnetite is sent by rail to Oxelösund for both smelting and export, whilst the rich haematite is chiefly smelted in Sweden. The main iron and steel manufacturing centre is Domnarvet: iron *mining* centres are underlined on the map. Many Swedish steel mills specialise in producing high-quality alloy steels. Simiiarly the metal-working industries make high-grade engineering products designed essentially for export. Some typical examples are:—

electrical machinery	cutlery
cutting tools	ball-bearings
armaments	Primus stoves
machine tools	dairy equipment
telephone equipment	

Other important industries include:—

TEXTILES—e.g. fine cottons at Göteborg and woollens at Norrköping

CHEMICALS esp. acids, explosives and fertilisers, e.g. at Norrköping.

SHIPBUILDING—at Göteborg and Malmö (S. Sweden)

WOOD-USING INDUSTRIES—See p. 42.

Old waterfront, Stockholm.

Mounting large roller-bearings in a Stockholm engineering works.

also more open than Northern Sweden to the westerly winds ((*23*) *Why?*) and so its rainfall of some 20″ is well distributed through the year. Thus grass grows in most months and the region is renowned for its dairy farms. Along the coast are many small market towns and industrial centres, and Trelleborg, Hälsingborg and Malmö are ferry ports with links to Denmark and Germany. Malmö (380 000) is the third town of Sweden, with many industries including: cotton textiles, margarine, soap, oil refining and brewing. (*24*) Name the main raw materials used in each of these industries. Which of them are imported?

Sweden is one of the most prosperous countries in the world, her prosperity being based on the possession of rich natural resources. Unlike most other Europeans the hard-working, inventive Swedes have been able to develop their country's wealth unhindered by devastation, for Sweden has avoided warfare since 1815. Although 90% self-sufficient in foodstuffs, Sweden leans heavily on imports of such vital commodities as coal, petroleum, textile fibres and tropical produce. To pay for these supplies from overseas the Swedes rely on a booming export trade. (*25*) Name three main types of exports from Sweden. The fact that the two largest cities, Stockholm and Göteborg, are also the two main ports, emphasises the importance of foreign trade to Sweden.

STOCKHOLM	GÖTEBORG
Built on group of islands at entrance to Lake Malaren, and so sometimes called 'The Venice of the North'.	Sweden's second largest city and prime port, lying on an excellent deep-water harbour at western end of Central Lakes waterway.
The lowest bridging point on Lake Malaren and so a natural route centre. (26) Explain this by a simple sketch-map.	A natural outlet for trade from industrial towns in and around Lakes Vättern and Vänern, especially since construction in the early C19th. of the Gota Canal.
Terminus of land and water routes across the Central Lakes region. (27) Name the main waterway from the map opposite.	Supplied with hydro-electric power from Tröllhattan Falls.
Centrally placed on Sweden's Baltic coast and kept open all year by ice-breakers.	Ice-free throughout the year ·and favoured by a very small tidal range. ((29) *What advantage is this to shipping?*)
Situated in the most densely peopled part of Sweden: hence is an excellent distributing centre. (28) Explain.	Has grown into a major industrial city with many industries based on imported raw materials, e.g. sugar-refining, textiles. Is Sweden's main shipbuilding and marine engineering centre.
Sweden's largest industrial city, with important *METALLURGICAL* and *ENGINEERING* works. Also book-binding and printing, clothing manufacture and preparation of foodstuffs.	
Capital of Sweden and the main cultural, commercial, financial and entertainment centre.	

	Imports	Exports
	Coal	Paper
	Petroleum	Wood Pulp
	Steel goods	Timber
	Chemicals	Iron and Steel
	Textile fibres	Machinery

FINLAND lies in one of the most extraordinary parts of the world, for this scene of lakes and pine forests—a sparsely-peopled wilderness—is typical of most of the country. Look at the map: even on this small scale over 250 lakes are shown. In fact there are over 60 000, and the native name for Finland, *Suomi*, means 'Lakeland'. The empty appearance of most of Finland's beautiful, lake-studded countryside is due partly to its high latitude. ((*30*) Are any parts of Britain as far north as Helsinki?) This, coupled with its 'interior' position, gives Finland a severe continental climate (*p. 16*). Long, very cold winters are followed by warm, brief summers. (Helsinki: July 17°C.; Jan. −5°C.) Snow, equivalent to about 375 mm rainfall, rests on the ground for about seven months in the north, and over three months on the southern plains. But Finland remains unattractive to Man chiefly because it was severely glaciated. The great Scandinavian ice-sheet gouged hollows all over the surface of this ancient granite peneplain (*pp. 6 and 10*): huge areas were swept bare of soil, and elsewhere the land was strewn with mounds and ridges of sand, gravel and boulder-clay.

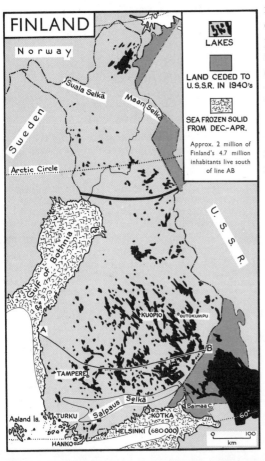

FINLAND

Norway

Suala Selkä

Maan Selkä

Sweden

Arctic Circle

Gulf of Bothnia

U.S.S.R.

KUOPIO   OUTOKUMPU

A

B

TAMPERE

Salpaus Selkä

Aaland Is.   TURKU

Saimaa C.

KOTKA

HELSINKI (680 000)

HANKO

0   100

km

LAKES

LAND CEDED TO U.S.S.R. IN 1940's

SEA FROZEN SOLID FROM DEC.-APR.

Approx. 2 million of Finland's 4.7 million inhabitants live south of line AB

The Saimaa Canal, passing through Soviet territory, is used to transport forest products to the coast.

NORTHERN FINLAND is one of the few parts of Europe still in the pioneer stage of development. This 'frontier' third of the country, lying north of the heavy line on the map, contains much almost uninhabited territory, where thousands of would-be farmers are fighting the elements to carve farms for themselves out of virgin forest. Many of these pioneers formerly lived in areas, shaded on the map, which were absorbed into the Soviet Union after the Second World War.

> ". . . Today they are still struggling for existence. . . . The forest, which covers almost the whole area, must be laboriously cleared by human effort. There is no money for bulldozers. Timber yields little or nothing because marshland produces only poor quality wood and, in any case, the farmer cannot afford to pay for its transport to the nearest mills."*

In spite of the harsh climate, poor soils and extensive swamps, these pioneers rear cattle and grow barley, oats and potatoes up to 200 kilometres north of the Arctic Circle. Aided by government subsidies, some of them even send dairy produce south to Finland's main towns. Farther north, however, beyond the low ranges of Maan Selkä and Suala Selkä, lie forests where the only inhabitants are nomadic Lapps (*see overleaf*).

THE LAKES PLATEAU resembles Northern Finland, for it too is an undulating peneplain, densely covered by pine and spruce forests. But the climate, though still severe, is less daunting, and farmers have penetrated into the Lakes region for centuries. Thus the forest is dotted with farm clearings; but the soils are thin and infertile except on occasional patches of boulder-clay or reclaimed swamp. Hay, oats and potatoes are the main crops, and almost every farmer has some dairy cattle.

Forestry, however, is the main source of livelihood, and most farmers take up lumbering in the winter months. Forest still covers 64% of the countryside, and lumber and timber products make up 57% of Finnish exports. The pattern of felling, transporting and processing the timber is similar to that in Sweden (*p. 35*); (*31*) account for the distribution of saw-mills shown on this map.

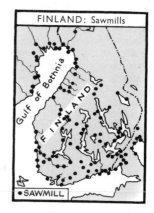

FINLAND: Sawmills

• SAWMILL

The southern rim of the Lakes Plateau is marked by an enormous terminal moraine called the Salpaus Selkä. Many of the streams which plunge southwards over the edge of this moraine have been harnessed to produce hydro-electricity.

* Wendy Hall, 'Resettlement in Finland',
Geographical Magazine.

FINLAND: Lines indicate length of growing season in days

Power is thus available for saw-mill and timber-working centres such as Tampere and Kuopio. Other industries in these towns are mentioned in the notes below. The Lakes Plateau also contains Finland's only important mining area, at Outokumpu, where copper, zinc and cobalt are mined.

THE COAST LOW-LANDS. (*32*) Study the map on p. 46 and complete the following sentence:—'*The region south of Tampere covers only* ...% *of the country but* ...% *of the total population live there: indeed* ...% *live in Helsinki, the capital.*'

One reason for this concentration of people in the south is its pleasanter climate. (*33*) How much longer is the growing season in Uudenmaa than in Kuopio? (*34*) How much warmer is Helsinki in July and January than Kuopio in corresponding months? What are the mean annual temperatures at each of these places? The coast plains, both in the south and along the Gulf of Bothnia also have the advantage of containing the best soils in the country. These occur on patches of clay which have emerged from below sea-level during the last million years. Thus,

---

FINLAND: MAIN TOWNS AND INDUSTRIES

TAMPERE
Chief industrial town.
Large H.E.P. stations.
Cotton mills, saw-mills, wood-pulp and paper mills, boot and shoe factories. Engineering, e.g. machines for paper and timber mills, transport equipment. New industries include jewelry, furs, saunabaths and chemicals made from imported petroleum. Clothing factories in Tampere and LAHTI produce c. 6% of Finnish exports.

HELSINKI
Capital of Finland.
Good, sheltered harbour.
Largest and best equipped port in the country.
A major distribution and market centre.
Timber, pulp and paper mills, textile and clothing factories, food processing works (e.g. flour mills), engineering works and shipbuilding yards. Business, cultural and entertainments centre.

HANKO
Exports much dairy produce.

KOTKA
Finland's largest timber port.
Wood-pulp and paper mills, large saw-mills.

TURKU
Main port for trade with Sweden.
Shipbuilding and marine engineering yards, ironworks.

(*35*) *Mark these towns and their industries on a sketch map.*

Finland's main growth industry is *engineering*. Exports of goods such as diesel engines, electronic gear and ships (notably icebreakers and luxury cruisers from Helsinki yards) are gradually lessening the country's traditional dependence on lumbering.

although they are still 70% forested, the plains are the main farm-ing area in Finland. Farming is very like that in the Central Lakes region (*p. 47*) except that rye and wheat are more impor-tant, and hay yields are heavier. ((*36*) Suggest reasons for these differences). In recent years dairying has become widespread and most farmers now send milk daily to co-operative creameries for butter production. Fishing is important in the Åland Islands.

THE LAPPS are nomadic reindeer herders who live in the extreme north of Norway, Sweden, Finland and the Kola Penin-sula of the U.S.S.R. (These countries contain about 19 000, 7700, 1800 and 1700 Lapps respectively.) The reindeer, which are semi-domesticated, provide them with meat, milk and skins, for clothing and bedding, and are used as draught animals to pull sledges. The Lapps follow their herds as they browse across the Arctic wastes for their main food, lichens. In summer they feed on poor mountain pasture, but in winter they migrate south-eastwards to shelter in the coniferous forests on lower ground. There is an annual spring round-up of the reindeer.

ICELAND has had links with Scandinavia for hundreds of years. Most of the island's 200 000 inhabitants are descendants of Viking settlers from the 9th century onwards. In modern times Iceland had political associations with Denmark, but in 1944 it became an independent republic.

Iceland is mountainous, barren, infertile and very sparsely in-habited. In glacial times it was heavily eroded by an ice-cap which completely covered the island: remnants of this ice-cap still lie over the high ground. The island consists entirely of volcanic rocks and there are still signs of vulcanicity, e.g. hot-springs, geysers and mud-flows. Most of the lava soil is very permeable, and so virtually useless for cultivation. Farming is also limited by the extensive morainic sands and by the climate, which is cool and wet. (*37*) Explain why, in spite of its sub-Arctic latitude, Iceland does *not* have a tundra climate (*pp. 14 and 15*). The main farming activity is sheep-rearing: 800,000 sheep provide wool, skins and frozen meat for export and wool for local clothing and carpet factories. The main crops are hay and potatoes.

Fishing is the leading industry, based on Reykjavik (91 000), the only large town. Cod, haddock, herring and fish products (e.g. fertilisers) form over 90% of the country's exports. The main imports consist of fuel oil and foodstuffs.

Sandy heathland, Blaavand, W. Jutland.

Terminal moraines, central Jutland.

| N. Frisian Is. | | | Dunes | Outwash Gravels | | Termina |
| N. Sea | ↓ | Lagoon Coast | Planted with conifers and marram grass | West Jutland | | Some barle |

x

50

DENMARK is a small country with a population of five million, i.e. less than half that of London. Your atlas relief map will suggest that it is rather monotonously flat—very little of the land is higher than 70 m. Yet within Denmark's small area there are remarkable contrasts in landscape and land use. These differences are due to the variety of glacial 'drift' deposits which almost cover the country. Look carefully at these photographs, maps and diagrams and (38) write the following paragraph correctly:—

Three distinct types of country are found in Denmark. I. In $\frac{\text{Western}}{\text{Eastern}}$ Jutland the ground surface is covered with glacial $\frac{\text{sands and gravels}}{\text{boulder-clay}}$, dropped there by melt-waters flowing $\frac{\text{eastwards}}{\text{westwards}}$ from the great $\frac{\text{lateral}}{\text{terminal}}$ moraines which run $\frac{\text{east–west}}{\text{north–south}}$ through the $\frac{\text{centre}}{\text{east}}$ of the Peninsula. The sandy soils in Western Jutland are very $\frac{\text{infertile}}{\text{fertile}}$. Much of the land

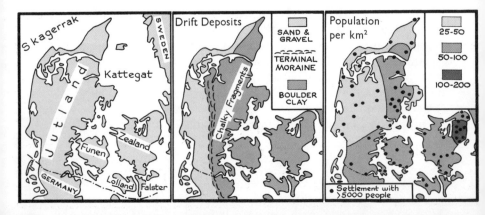

Boulder clay country, Zealand

Moraine & oats grown

Boulder Clay of East Jutland & Danish Islands

Baltic Sea

was formerly heath and bog, and despite large-scale reclamation this part of Denmark still has the lowest proportion of $\frac{pastoral}{arable}$ land. The $\frac{west}{east}$ coast is fringed with dunes and tidal marshes. To prevent sand being blown inland large areas have been planted with c....... and m..... g..... 2. The morainic hills of central Jutland also contain much $\frac{infertile}{fertile}$ stone, gravel and sand. Although the hummocky ridges are awkward to cultivate, fodder crops such as b....., r.. and o... are grown, and this part of Denmark is classified as $\frac{PASTURE+crop}{CROP+pasture}$ land. 3. Eastern Jutland and the Islands consist of $\frac{ground\ moraine}{outwash\ gravels}$. The boulder clays here are $\frac{moderately}{extremely}$ fertile. As they contain fragments of $\frac{chalk}{granite}$ they are also limey, light and $\frac{well}{little}$ suited to cultivation. This is the main $\frac{arable}{pastoral}$ farming region in Denmark.

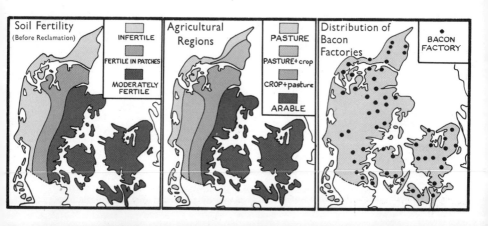

Soil Fertility
(Before Reclamation)

INFERTILE

FERTILE IN PATCHES

MODERATELY FERTILE

Agricultural Regions

PASTURE

PASTURE+ crop

CROP+pasture

ARABLE

Distribution of Bacon Factories

BACON FACTORY

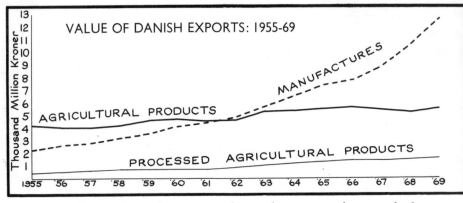

VALUE OF DANISH EXPORTS: 1955-69

AGRICULTURAL PRODUCTS

MANUFACTURES

PROCESSED AGRICULTURAL PRODUCTS

Denmark produces no coal, metal ores, petroleum or hydro-electricity. Over wide areas her soils, in their natural state, are peaty or sandy and infertile. In spite of these drawbacks the Danes make remarkable use of their limited resources and enjoy a high standard of living. Until very recently this prosperity depended, to an almost excessive extent, on Denmark's famous scientific farming, with its emphasis on dairying and pig-rearing. Who, indeed, has not heard of Danish butter or Danish bacon? Yet the chart shows that significant changes are taking place in the Danish economy and these trends are certain to continue.*

The main reason for this relative decline of agriculture is the difficulty Denmark now finds in selling so much dairy produce to other European countries. Even so, farm products still form such a large fraction of Danish exports that the continued prosperity of her agriculture is essential.

Dairy farming was of little importance in Denmark until about 1870. Before that date most Danish farmers grew cereals, and Denmark exported grain. Then came a flood of cheap grain imports to Europe from the newly developed prairies of North America and, faced with such keen competition, many farmers switched to dairying. The table shows how remarkably successful they have been. Danish dairy farmers have been so successful largely because of the farming **Co-operative**

FACTS ABOUT DANISH DAIRYING

	1871	1969
Total number of dairy cattle	808 000	1 233 000
Milk yield per cow	1 tonne	4.5 tonnes
Total output of butter	39 000 tonnes	144 000 tonnes

Total world trade in butter............ c. 600 000 tonnes p.a.
Total exports of Danish butter............ c. 100 000 tonnes p.a.

(39) *What was the percentage increase between 1871 and 1969 in (a) the number of dairy cows; (b) the annual yield of milk and (c) the total output of butter?* (40) *Estimate the fraction of the world's butter trade controlled by Denmark.*

* "In the future, emphasis will be increasingly on manufacturing industry, and it will be expedient to work for the transfer of labour from agriculture to urban industries." Prof. Kjeld Philip, writing in the *Financial Times*. In fact the high wages offered in Danish factories *are* luring large numbers of farmworkers off the land and Danish farming is changing as a result. Since 1960 over 5 000 small farms, unable to offer attractive wages or to afford expensive machinery,

Associations. These voluntary organisations were formed (*a*) to collect milk from farmers and take it to conveniently located factories for processing and (*b*) to get the butter, cream and cheese quickly and cheaply to markets in Denmark and overseas. Nowadays there are Co-operatives for nearly every aspect of dairy-farming, including bacon-curing, the purchase of fodder and fertilisers, testing the quality of cattle and milk and making loans.

In spite of the poor soils over large parts of Denmark (*39*) *estimate from the map on p. 51 the proportion of the country classified as naturally infertile*) the farmers have made tremendous efforts to improve their land by draining and liming waterlogged clays, reclaiming peaty heathland and adding abundant fertilisers and animal manures. As a result 90% of the land surface is productive farmland, and yields of fodder crops such as barley, oats, sugar-beet and hay are amongst the highest in Europe. Danish cattle, too, are of the highest quality that scientific breeding can produce. As the cows graze they are moved methodically over every blade of grass and clover by single-strand portable electrified fences. In addition they are stall-fed with enriching cattle-cake concentrates, made from imported oil-seeds and maize.

Pig-rearing in Denmark goes hand in hand with dairying, for the staple food of the pigs is skimmed milk, butter milk and whey. These 'by-products' are sold back to the farmers by the Co-operative factories after the cream has been separated from the fresh milk. The pigs are stall-fed in hygienic, well-lit sheds, and great care is taken to breed and rear high quality animals. The Danish 'Landrace' variety of pig has been bred to give long sides, good hams and light shoulders. Sixty co-operative bacon factories handle some 9·5 million pigs annually, and 60% of the pig meat they produce is exported, mostly to Great Briton. No part of the pig is wasted and by-products from the factories include lard, bone and blood meal and bristles for brushes.

have been forced to close, and deserted farm houses have become commonplace. The typical modern Danish farm is about 20 hectares and the land is used solely for grazing and the intensive cultivation of grasses and animal feed-grains, especially barley. Dairying is gradually giving way to pig and poultry rearing, due partly to increases in the costs of milk production and also to an acute shortage of skilled dairymen. There is also a growing export market in Europe for pig-meats, veal, beef, mutton, eggs and poultry. Most Danish livestock is now permanently housed, the animals being fed with grass briquettes and fodder pellets. Many Danish farmers obtain additional income by working part-time in local factories and by letting out summer cottages to holiday makers. Another modern trend is an increase in intensive market gardening, especially around Copenhagen.

MAIN MANUFACTURERS IN DENMARK

From Danish Farm Produce	From Imported Metals	From Imported Oil-seeds	Based on Fishing	Other Manufactures

Butter, fish fertilisers, diesel engines, beer, dairy equipment, frozen fish fillets, transport equipment, clothing, vegetable oils, herring-oil preparations, drugs, cattle-cake, cement, offal-fertilisers, canned fish, beet sugar, condensed milk, bacon, ships, cheese, cement-making machinery, margarine.

Industry in Denmark. Cheese and fertilisers, margarine and diesel engines: these are a few of the great variety of products manufactured in modern Denmark. (*40*) Look carefully through the fuller list given above and re-arrange the articles in the tabular form suggested. You will then see that these miscellaneous goods are not quite so unrelated as one might at first imagine.

Copenhagen (København or Merchants' Harbour) is the only large centre of industry in Denmark. It is also the leading port and the main centre of administration, business, commerce and culture. Copenhagen occupies a very important strategic location: (*41*) study its position on an atlas map and suggest why it dominates (and for hundreds of years controlled) trade in the Baltic Sea. Copenhagen grew up originally as a small fishing port on a deep, sheltered harbour, but in medieval times it blossomed into a great commercial centre, military and naval base and entrepôt for most of Northern Europe. ((*42*) *Explain 'entrepôt'*). Copenhagen's dominance of the Baltic trade was challenged in 1895 by the opening of the Kiel Canal ((*43*) *suggest why*), but by creating a 'free port' with very favourable customs and cargo-handling dues Copenhagen so encouraged the growth of her

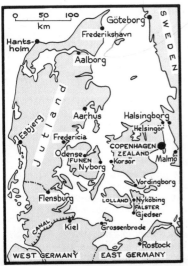

industry and trade that she remains pre-eminent among Baltic ports. Today Copenhagen handles about 60% of Danish imports but only about 33% of the country's exports. This discrepancy is explained by the vital importance of the trade in dairy products and pig meat with the U.K. To handle this trade an entirely artificial port was built in 1874 at *Esbjerg* on the west coast of Jutland. ((*44*) *Why there?*) Fast diesel trains link Esbjerg to Copenhagen. Esbjerg is also Denmark's

SOME IMPORTANT INDUSTRIES	TRADE	
Metallurgy (from imported ores and metals). Shipbuilding and engineering (mostly making machines, transport and electrical goods). Textiles, clothing and leather manufacture. Making pottery and porcelain (from kaolin dug on the Danish island of Bornholm). Food and fertiliser industries (see list on opposite page).	*Imports* textiles machinery fuel & oil fodder fertilisers foodstuffs	*Exports* dairy produce meat live animals machinery transport equipment eggs

principal fishing port. In 1969 an important new artificial port was opened at Hantsholm, with trading, fishing and refrigerating facilities and ferry links to England, Scotland and Norway.

Apart from the capital there are few large towns in Denmark, for one in every four Danes depends directly on agriculture for a living and so lives in a village, hamlet or isolated farm. The second largest town is *Aarhus* (population 190 000), an important marketing centre in Eastern Jutland. Another smaller market town is *Odense*. *Aalborg* in Northern Jutland is famous for its cement, produced from the chalk underlying the town. The world's largest rotary cement kiln is located there.

COMMUNICATIONS IN DENMARK are obviously difficult, but the problem has been met by a remarkable series of ferries and bridges. The most important are listed below. (45) Show them on a large sketch-map (your atlas and the map on p. 54 will help):

1. Rail routes enter Denmark via Hamburg and Flensburg.
2. A great road and rail bridge (see photo) links Jutland to Funen across the Little Belt at Fredericia.
3. A train and vehicle ferry links Funen and Zealand across the Great Belt.
4. Bridges at Vordingborg and Nykobing link Zealand, Falster and Lolland.
5. The following train and vehicle ferries link Denmark to Sweden and Germany: Frederiskshavn–Göteborg; Gjedser–Rostock; Gjedser–Grossenbrode; Helsingör–Halsingborg; Copenhagen–Malmö.

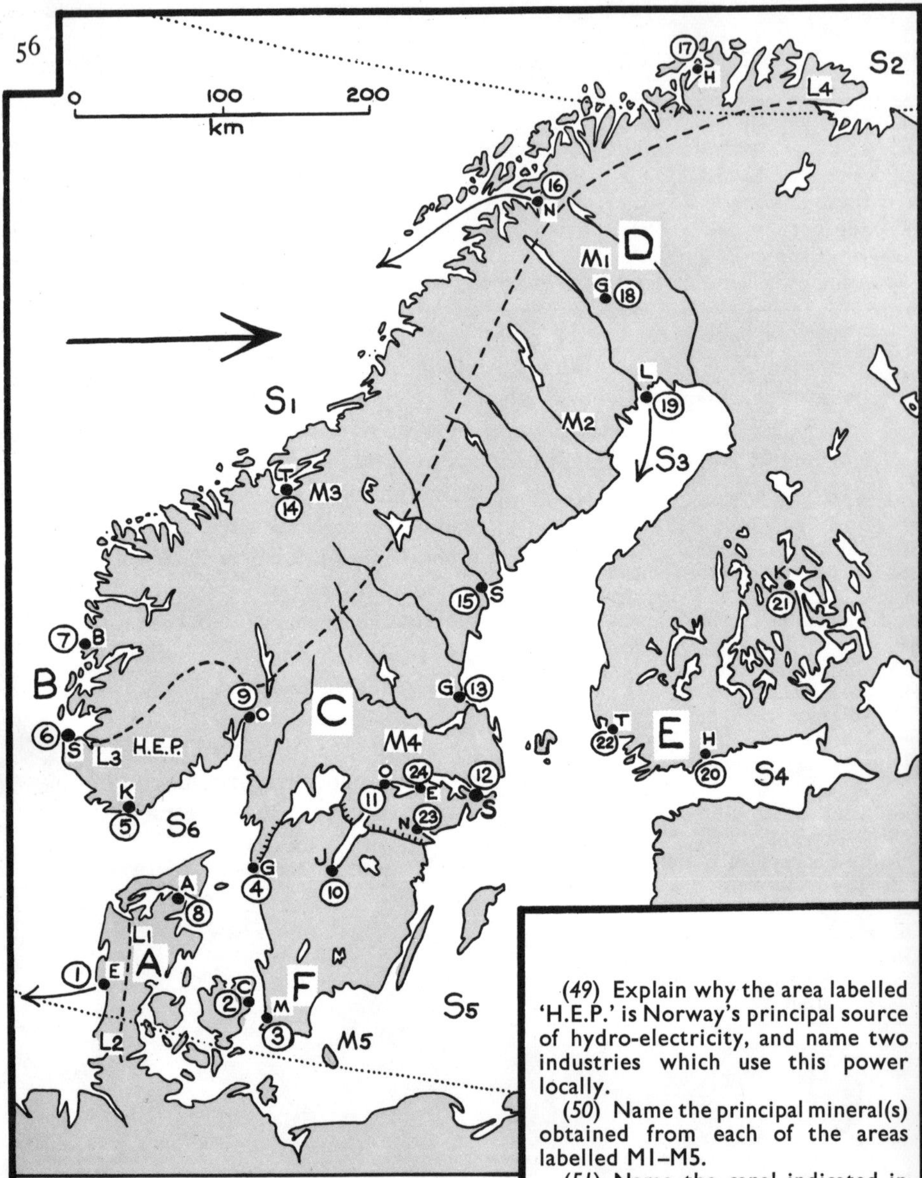

100   200
km

17
H

S2

L4

16
N

M1
G   18
D

L
19

M2

S3

S1

T
14   M3

21
K

15   S

G   13

7   B

B

9

C

M4

G
22   T

E
H

6   S
L3   H.E.P.

24   O
11   E   12
S

20
S4

K
5   S6

N   23

G
4

J
10

A
8

M

L1
E   A

F
M

1

2   C

3   M
S5

L2

M5

Test Map of
SCANDINAVIA

(49) Explain why the area labelled 'H.E.P.' is Norway's principal source of hydro-electricity, and name two industries which use this power locally.

(50) Name the principal mineral(s) obtained from each of the areas labelled M1–M5.

(51) Name the canal indicated in central Sweden and discuss its economic importance.

(52) Describe the course of a typical river in the region between C and M1, showing its relation to the economic life of the countryside.

(53) What is the significance of each of the arrows drawn on the map?

(54) Describe (a) the landscape, (b) the climate and (c) the main economic activities in each of the areas labelled A–F.

(46) Name the two lines of latitude drawn on the map.

(47) Name the sea areas S1–S6 and state (a) which areas are frozen over in winter, and (b) which areas are important fishing grounds.

(48) What is the significance of the broken lines L1–L2 and L3–4?

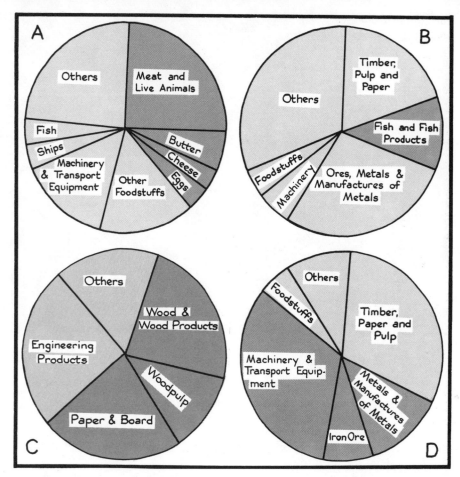

These four diagrams show in detail the *export* trade of Norway, Sweden, Denmark and Finland, *but not in that order*. (57) Identify the country to which each diagram applies; the shaded portion in each case gives a positive clue. Check your answers with the correct order given at the foot of this page.

There are no such distinctive differences between the main *imports* of these countries; in each case they are as follows:— *machinery and transport equipment, metals and metal goods, fuel and oil, textiles and textile raw materials, foodstuffs.* (58) Suggest geographical reasons for this pattern.

All these countries, too, conduct the bulk of their overseas trade (65%–75% in each case) with member countries of either E.E.C. (the Common Market) or E.F.T.A. (*p. 31*). The only notable exception is Finland, which has about 17% of its foreign trade with Comecon countries. (59) Suggest why.

Paris, Île de la Cité (see p. 63).

CHAPTER 3

France

—is the largest country in Europe, except Russia.

—is about twice as big as the British Isles.

—occupies a prominent political position in Western Europe.

—has coastlines on three sea areas ((*1*) *name them*) each crossed by vital trade routes.

—has a vulnerable north-east frontier bordered by Belgium, Luxembourg and Germany. (*2*) *Where does the relief map suggest this frontier is most vulnerable?* (*3*) *Which country has invaded France three times since 1871?*

—was formerly a colonial power, but has now granted independence to nearly all of its overseas possessions. (*4*) *Find out which territories France has given up since 1949 in* (a) *Asia and* (b) *Africa.*

This map shows the ten main geographical regions into which France may be divided. Look again at the map on page 8 and at the relief map in your atlas and (*5*) name the three regions of France which are very mountainous. (*6*) Which two of these highland regions are composed of young fold ranges? (*7*) What is the geological history of the third mountainous region? (*8*) What do the worn-down ranges of the Brittany Peninsula have in common with the Vosges and the Ardennes Plateau?

58

FRANCE: Natural Regions, Rivers and Power Resources

The large French island of Corsica consists mainly of hard crystalline rocks. It is largely mountainous & rugged. Only 2% of the island is cultivated. Goat- & sheep- rearing provide the main source of livelihood for the 280,000 population. An expanding tourist industry is a welcome additional source of income.

COALFIELDS

IRON ORE

PETROLEUM

NATURAL GAS

NUCLEAR POWER STN.

H.E.P. STNS.: MAIN AREAS ONLY
(Symbols roughly proportional to installed capacity)

(9) Name the rivers labelled 1–15 on the map. Which one, for part of its course, forms the frontier between France and Germany?

(10) What natural gaps or corridors link the following pairs of regions:— Aquitaine Basin and Mediterranean France? Aquitaine Basin and Paris Basin? Paris Basin and Mediterranean France?

(11) Name the gap linking the valleys of the upper Saône and the Rhine.

Broadly speaking, France experiences three distinct types of climate:—(i) mild and equable, moist in all seasons; (ii) hot, dry summers with a drought and warm, wet winters; (iii) cold winters, hot sum-mers, with most rain falling in summer.

(12) Read pages 14–18 again and state, with reasons, which climate you would expect in each of the following places:— Brest, Marseilles, Nancy.

(13) From the map on page 11 state which regions have substantial surface deposits of fertile *limon* (loess).

(14) Which region contains the largest coalfield in France?

(15) Which other region has small, scattered coal deposits?

(16) Name two important mineral deposits in Alsace-Lorraine.

(17) Which regions contain important hydro-electric power stations?

Wheatfield, Beauce.

Armorica | Hills of Perche | Ile de France | PARIS

CHALK

LIMESTONE

MOSTLY SANDS &

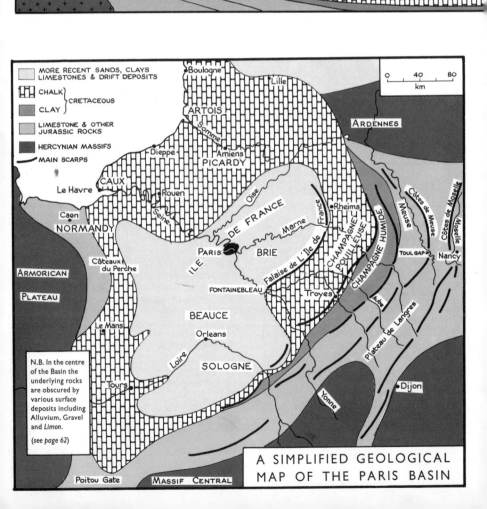

A SIMPLIFIED GEOLOGICAL MAP OF THE PARIS BASIN

MORE RECENT SANDS, CLAYS LIMESTONES & DRIFT DEPOSITS

CHALK — CRETACEOUS

CLAY

LIMESTONE & OTHER JURASSIC ROCKS

HERCYNIAN MASSIFS

MAIN SCARPS

N.B. In the centre of the Basin the underlying rocks are obscured by various surface deposits including Alluvium, Gravel and *Limon*.

(see page 62)

0 40 80
km

Boulogne
Lille
ARTOIS
ARDENNES
Somme
Dieppe
Amiens
PICARDY
CAUX
Le Havre
Rouen
Oise
ILE DE FRANCE
Rheims
Côtes de Moselle
Seine
Marne
CHAMPAGNE
POUILLEUSE
Meuse
Caen
NORMANDY
Paris
BRIE
Falaise de l'Ile de France
CHAMPAGNE HUMIDE
TOUL GAP
Nancy
Côteaux du Perche
FONTAINEBLEAU
Troyes
Aube
ARMORICAN
PLATEAU
BEAUCE
Plateau de Langres
Le Mans
Orleans
Loire
SOLOGNE
Tours
Yonne
Dijon
Poitou Gate
MASSIF CENTRAL

Vineyards near Epernay.

Pasture land, Champagne Humide.

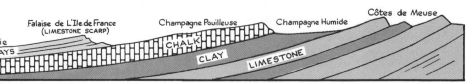

Falaise de L'Ile de France (LIMESTONE SCARP)   Champagne Pouilleuse   Champagne Humide   Côtes de Meuse

CHALK   CLAY   LIMESTONE

THE PARIS BASIN, besides containing the capital, is a highly important agricultural region. These maps, photographs and diagrams show the main facts about its geography. Study them carefully and (*18*) complete the following passage correctly:—

The Paris Basin is an almost circular lowland. In the $\frac{\text{north and east}}{\text{south and west}}$ this lowland is drained by the River Somme and by the River Seine and its tributaries, the most important of which are the A..., Y...., O... and M....; the $\frac{\text{south-western}}{\text{north-eastern}}$ portion is drained by the River L..... Structurally the Basin is a great $\frac{\text{anticline}}{\text{syncline}}$ or hollow, formed by successive layers of rock which lie within one another like saucers. The lowest 'saucer' is made of $\frac{\text{limestone}}{\text{clay}}$; at the edges of the Basin this same rock outcrops to form a broken rim of higher ground, as in the Côteaux du P..... to the $\frac{\text{east}}{\text{west}}$ and the Côtes de M...... to the $\frac{\text{east}}{\text{west}}$. Other

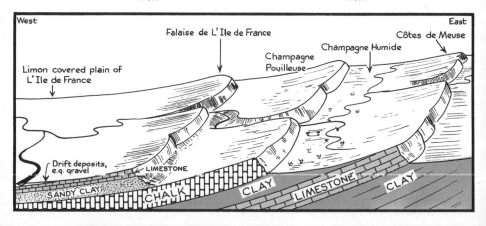

West   East

Falaise de L' Ile de France   Côtes de Meuse

Limon covered plain of L' Ile de France   Champagne Pouilleuse   Champagne Humide

Drift deposits, e.g. gravel   LIMESTONE

SANDY CLAY   CHALK   CLAY   LIMESTONE   CLAY

'saucer rims' are represented by the $\frac{limestone}{chalk}$ escarpment of the Île de France. Sandwiched between the chalk and limestones are layers of $\frac{sandstone}{clay}$; where this rock occurs at the surface there are $\frac{moist\ vales}{dry\ uplands}$ such as the $\frac{Pouilleuse}{Humide}$ district of Champagne. The chalk 'saucer' contains various $\frac{drift\ deposits}{rocks}$, e.g. clay, sandstone or limestone, and $\frac{drift\ deposits}{rocks}$ such as river gravels, *limon* and alluvium, which give rise to minor ridges and varied scenery immediately around Paris.

These different types of rock and drift deposit greatly affect the mode of life of people within the Paris Basin.   Details about farming in the main sub-regions are given in the notes below.   (*19*)

MAIN SUB-REGIONS OF THE PARIS BASIN

Île de France (the Island of France). Inner 'heart' of the Basin. Outer limits clearly demarcated by marked limestone scarp, i.e. the Falaise de l'Île de France.   Scenery and farming vary according to surface deposits and fertility. FARMING with emphasis on DAIRYING.   Many farm hamlets.

Beauce. Limestone plateau covered by *limon*.   Soils extremely fertile.   Large farms and large hedgeless fields.   Much WHEAT and SUGAR-BEET, with DAIRYING towards west.   Much MARKET GARDENING in Loire Valley downstream from Orleans.

Brie. Limestone at surface contains high percentage of clay.   Surface thus often wet, and drained by many streams. Rich, heavy soils.   Much DAIRYING to supply Paris.   WHEAT and SUGAR-BEET on the better drained soils.

Sologne. An infertile scantily peopled district underlain by sterile sands and lake-studded clays.

Fontainebleau. Very sandy and infertile. Much forest.

Champagne
(a) 'Champagne *pouilleuse*' ('*lousy*').   A chalk upland.   Little surface drainage. Thin infertile soils.   Many SHEEP. Little cultivation except for terraced VINEYARDS on south-facing scarps.

(b) '*Champagne Humide.*'   A broad, moist clay vale.   Rich pastures on heavy, impermeable soils.   MIXED

Normandy. Narrow bands of chalk, limestone and clay outcrop rather steeply against the old, hard rocks of Armorica.   Rapid alternation of rock types gives rise to a varied countryside and agriculture.   Cereals and roots, notably OATS and POTATOES, grow on *limon*-covered chalk.   DAIRYING (favoured by mild, moist climate), MARKET GARDENING and CIDER APPLE ORCHARDS in clay vales.

Northern Chalklands. A series of low, rolling chalk plateaus in Caux, Artois and Picardy.   Remarkably fertile due to thick covering of *limon*.   An important arable farming area, growing mainly WHEAT and SUGAR-BEET. Many cattle reared in Caux: emphasis on DAIRYING.   Much dairy produce sent to Paris and to coast towns, e.g. Rouen and Le Havre.

Eastern Limestone Scarplands. A series of limestone escarpments, separated by wet clay vales.   The escarpments become higher towards east and exceed 650 m in the Plateau de Langres.   Scarps are broken at intervals by river-cut gaps into isolated lengths called *Côtes*.   Higher ground used for rough grazing (SHEEP) or FORESTED.   Lower slopes cultivated. Many CATTLE in the vales.   Terraced VINEYARDS on south-facing scarps (*esp. the '*Côte d'Or*'—p. 71) produce BURGUNDY wine.

Use the notes to make a 'Farming Map of the Paris Basin' by adding appropriate labels to a *large* copy of the map on page 60.

PARIS gets its name from the *Parisii*, a Celtic tribe who lived on the island in the Seine now known as the Île de la Cité. (*(20) Why do you think they chose an island for their home?*) In Roman times Paris became an important crossing-point of the Seine for a military road leading northwards to the Channel ports. The names of Paris streets which still mark this old route are shown on Map **A**. To guard this bridging-point the Romans built a fortress and the first of a series of walls enclosing land on both banks of the river.

In medieval times Paris grew rapidly; its position at the heart of the richest farming region in France, and as a focus of many waterways, made it a natural market centre. It also became the home of the French monarchy and a centre of the Church (*(21) What famous cathedral is on the Île de la Cité?*) (*Photo p. 58.*)

The dominant position of Paris was reinforced in the late 18th century by the construction of a great system of roads, the *routes nationales*, which radiate from the city in all directions. The map on page 83 shows that Paris is also the focal point of the French railway network. The directions followed by the main road and rail routes from Paris are indicated on Map **B**. (*22*) Make a large copy and add the following labels in their appropriate places:—

To Calais, Boulogne and ferry services to England/To Dijon and Mediterranean France *via* Rhône–Saône Valley/To Cherbourg and trans-Atlantic liner service/To Orleans and Aquitaine *via* the Poitou Gate/To Brittany/To Rhine Valley towns, e.g. Basel and Strasbourg/To Rouen, Le Havre and Atlantic shipping routes/To the Industrial North-East.

Modern Paris and its adjacent suburbs form by far the largest

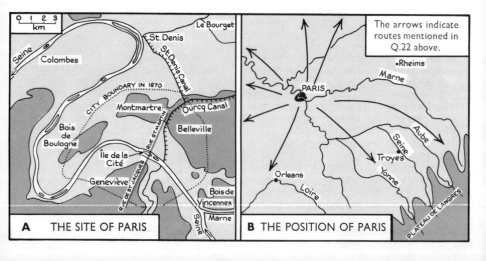

A THE SITE OF PARIS B THE POSITION OF PARIS

Main Industries of PARIS
Engineering, e.g.:— steel castings rolling stock machine tools armaments diesel engines Renault and Citroën cars
Consumer goods, e.g.:— women's luxury wear tobacco and cigarettes processed foods paper and books soap and cosmetics furniture footwear

conurbation in France, with a population of over 8½ million. In addition to being the capital it is also the main centre of French culture, entertainment, business and finance, and is the largest industrial centre in Continental Europe. Many bulky commodities such as coal, timber, steel, sugar and other foodstuffs reach Paris by barge, mostly upstream from Le Havre and Rouen. Thus many processing industries are located beside the Seine or in the eastern and north-eastern suburbs near the Ourcq and St. Denis Canals (*map* **A**, *p. 63*). Details of the more important of the enormous variety of industries in Paris are given on the left.

Other towns in the Paris Basin fall into two main categories:— (i) *market towns* and (ii) *ports*. Many of the market towns are located in gaps through the escarpments, or at points where route-ways cross the rivers. Examples include *Rheims*, the centre of the Champagne wine country; *Orleans*, on the great bend of the Loire; *Amiens*, the main market town for Artois and Picardy and *Troyes* on the upper Seine. Most of these towns have industries related in some way to agriculture. Such industries include the manufacture of agricultural implements and the processing of local agricultural products, e.g. flour-milling, sugar-refining and making biscuits. Textile manufacture is also widespread, having developed originally using wool from local sheep. Details of the two major ports in the Basin are given below. In addition there are the ferry ports of *Dieppe* and *Boulogne*, with regular sailings to Newhaven and Dover–Folkestone respectively.

ROUEN, on the Seine (*map p. 60*), has for centuries been the major port for the Paris Basin. It lies 80 miles from the open sea at the head of navigation for medium-sized ocean-going boats. As ships became larger it faced increasing competition from Le Havre, but is still the prime **outport** for Paris. About ¼ m. people live in and around Rouen. The maritime port extends downstream from the city for 19 km. Upstream lie 13 km of wharves for river barges.
Main imports: coal, petroleum,* cellulose, ores, foodstuffs.
Main industries: cotton textiles, metal smelting, chemicals, oil-refining, engineering.

*Pipelines carry petroleum from Le Havre to refineries near Rouen, and refined products from Rouen to Paris.

LE HAVRE is the main French liner port and chief commercial port engaged in freight trade across the North Atlantic. Sheltered in the mouth of the Seine Estuary, it has grown rapidly during the past century with the expansion of ocean passenger traffic. The city now has a population of some 200 000, and has spread from its original site on the alluvial plain up on to the adjacent chalk plateau. Development now in hand includes dredging deep-water berths to take the biggest ships afloat, and the reclaiming of 10 000 ha of land for factories.
Main imports: Petroleum,* cotton, timber, coal, oil-seeds.
Main industries: shipbuilding and repairing, engineering, flour-milling, oil-seed crushing, sugar-refining.

THE INDUSTRIAL NORTH-EAST (*French Flanders*), despite its small area, has been in some respects the most important region in modern France since it contains the country's main coalfield. The coal-bearing rocks lie well below the surface in the districts of Pas de Calais and Nord (*see map*), and continue eastwards into Belgium as the Sambre-Meuse coalfield (*pp. 126–128*). The coal is difficult and costly to work, for the rocks have been severely faulted and many seams are interrupted or are steeply inclined to the ground surface. Mining costs are increased, too, by the thinness of many seams. Even so this field is vitally important to France, for it produces more than half the total national coal output of about 47 million tonnes each year.

The main French colliery centres are Béthune, Lens, Douai and Valenciennes. Over one-half of the coal is utilised on the coalfield, mostly in coke-ovens, power stations or briquette works located at the pit-heads. Waste gases from the coke-ovens are fed into gas pipelines and distributed throughout the region for domestic and industrial heating. By-products from the cokeries have given rise also to a large chemical industry on the coalfield, especially at Béthune, Liéven and Courrières. Its varied products include drugs, fertilisers, dyes, plastics and detergents.

Coal-fired power stations supply electricity for many other industries in French Flanders. Foremost of these is textile manufacturing, for which this region has long been famous. The main textile centre is Lille, but the map shows that cloth is also produced in many other towns. ((*23*) *Name them.*) In medieval times local supplies of wool and flax were used to make woollens and linens. These are still produced, but today the main branch of the textile industry is cotton manufacturing. Other industries

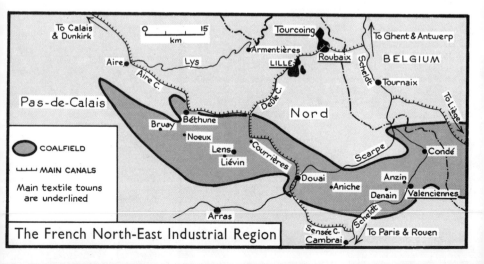

The French North-East Industrial Region

include iron and steel-making and engineering at Lille and Valenciennes. Iron ores from overseas and from Lorraine are smelted with local coke in both these towns. Engineering products include railway equipment and agricultural implements.

The main French ports of this region are Calais and Dunkirk, both of which are also ferry ports with links to Dover. Dunkirk is the more important; besides handling a large volume of transit trade it has food-processing industries and a huge oil-refinery. Much trade is also channelled through the Belgian port of Antwerp, linked with the French industrial towns by a network of broad canals and canalised rivers (*map p. 65*). Canals are the main form of transport for moving coal from the north-eastern field to other parts of France. Most shipments are to (i) adjacent industrial towns such as Lille and Roubaix, (ii) the iron and steel manufacturing region of Lorraine (*see below*) and (iii) Paris. The main coal loading centres are Lens and Douai.

(*24*) Use the information in the preceding four paragraphs to make a brief table of notes under the headings:—

MAIN INDUSTRIES IN NORTH-EAST FRANCE

Industries based on Coal	Textile Manufactures	Iron and Steel and Engineering	Port Industries

ALSACE-LORRAINE, too, contains highly important mineral deposits which have given rise to industrial activities. (*25*) What percentage of the total French output of each commodity listed beside the map is produced in Alsace-Lorraine?

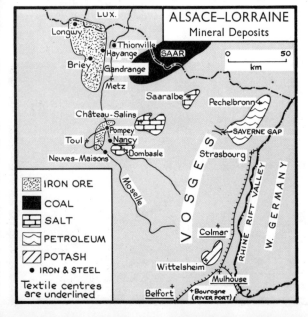

ALSACE–LORRAINE
Mineral Deposits

IRON ORE
COAL
SALT
PETROLEUM
POTASH
• IRON & STEEL
Textile centres are underlined

ANNUAL OUTPUT OF CERTAIN COMMODITIES
(Million Tonnes)

Commodity	Alsace-Lorraine	France
Iron Ore	52·3	55·2
Pig Iron	11·7	16·4
Crude steel	12·8	20·4
Potash	1·8	1·8
Salt	1·5	4·1
Coal	9·8	45·1
Petroleum	·07	2·7

Most important are the 'minette' iron ores of Lorraine. These occur at or near the surface in the limestones of the Côtes de Moselle. This is the largest iron ore field in Europe, but it has only been exploited since 1879, when for the first time it became possible to remove the phosphorus impurities. The iron content of the Lorraine ores is fairly low—between 25% and 35%. Although much ore is sent to Luxembourg, Belgium and the Saar, it is most profitable to smelt it in Lorraine, using coking coal brought from the Ruhr via the Moselle Canal (*see also p. 92*). Thus a major iron and steel manufacturing region has developed. (*26*) Name the main centres from the map. These towns contain many modern integrated iron and steel mills.

Other industries in this region include engineering, e.g. a Citroen factory at Metz makes gearboxes and other engine parts and gives employment to more than 4000 workers, chemicals and textile manufacturing. The latter is an old-established industry in Eastern France. The main centres are indicated on the map: power is obtained from hydro-electric stations in the nearby Vosges. Bleaches for the textile trade are one of the products derived from the salt deposits around Dombasle. Some of the salt is dug from mines, but mostly it is pumped from below ground as brine. Both the salt and the potash quarried near Mulhouse are used in the manufacture of chemicals.

Some parts of Eastern France support prosperous farm communities. Dairying and mixed arable farming are important in the clay vales between the limestone Côtes of Lorraine, and specialised crops are grown in the fertile parts of the Rhine Rift Valley in Alsace. (*27*) Use the labelled sketch below to write notes on land-use in this latter district.

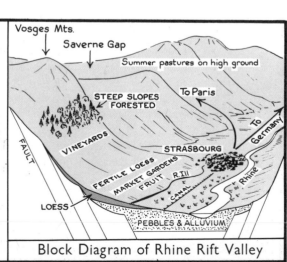

STRASBOURG
—the main market centre for Alsace, is a major river port and industrial town. Products based on local farming include flour, wine and tobacco.

Other industries include engineering, chemicals, textiles and oil-refining. Strasbourg lies on the oil pipeline linking Marseille to Karlsruhe. (*28*) Show by means of a labelled sketch-map that Strasbourg is well placed for trade with Lorraine and the Paris Basin *via* the Saverne and Toul Gaps, and with Germany *via* the Rhine Valley.

Block Diagram of Rhine Rift Valley

Gironde Estuary petroleum terminal.

A   Beach & Dunes   Lagoons   Landes   Gironde
                                                                 ↓

THE COAST OF SOUTH AQUITAINE is remarkably straight and unvaried. From Biarritz to the Gironde estuary stretch mile after mile of wide, gently shelving beaches, backed by lines of sand-dunes. The prevailing westerly winds carry sand far inland, and the desert-like region thus created is called the *Landes*. To prevent further encroachment of sand over fertile farmland, large areas of the Landes were planted with marram grass and pine trees. The plant roots bound the sand grains together and gradually stabilised the dunes beneath a permanent cover of vegetation.

The pine forests yield resin, tar, timber and wood-pulp and give employment to several thousand lumbermen. The 'Expansion Centre' aims to encourage more agriculture in Les Landes . . . " re-establishing the former balance between crops, cattle and forestry, a balance destroyed by the monster-like development of forestry as it invaded cultivated clearings, drove out the inhabitants, and by its excessive density created a permanent risk of fire."*

THE GIRONDE ESTUARY is the main 'funnel' for river drainage in Aquitaine. Heavy winter rainfall brings a constant danger of flooding, and vast quantities of silts, sands and gravels have been dumped in the lower valleys of the Garonne and the Dordogne and on the estuarine plains.

At the head of the Gironde, 96 km from the sea, stands the great trading port and regional capital of *Bordeaux*. Exports from Bordeaux are mainly timber, wines and

THE AQUITAINE BASIN is a triangular lowland bordered by the Bay of Biscay, the Pyrenees and the Massif Central. Its gently undulating surface is covered mainly by geologically recent sands, clays and silts, on which there are rich, easily cultivated soils. The climate is the most pleasant in France, with mild winters, hot summers and some rain in all seasons.

For over a thousand years people here have followed a rural and agricultural way of life, and as the region had no coal or iron deposits it derived little benefit from the Industrial Revolution. Being rather isolated from the rest of France it became unduly dependent for its prosperity on trade in two commodities, wine and timber. Recently, however, petroleum has been discovered

68

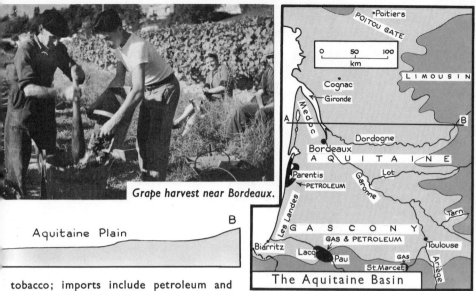

Grape harvest near Bordeaux.

Aquitaine Plain

The Aquitaine Basin

(Map labels: Poitiers, POITOU GATE, km, Cognac, Gironde, LIMOUSIN, Medoc, Dordogne, Bordeaux, AQUITAINE, Parentis, PETROLEUM, Lot, Garonne, Tarn, Les Landes, GASCONY, GAS & PETROLEUM, Biarritz, Lacq, Pau, Toulouse, GAS, St.Marcet, Ariège)

tobacco; imports include petroleum and tropical commodities such as vegetable oils and cane sugar for refining. The old-established wine trade of Bordeaux is based on high-quality wines and brandies produced in the local region.

The rich alluvial soils of the valley terraces are intensively cultivated, with large numbers of vineyards, orchards and market, gardens. Cereal crops such as wheat, barley and maize are also grown throughout Aquitaine, and in the valleys in the north-east dairying is a speciality. The fertile soils and warm, moist summers also favour tobacco cultivation, and this crop is produced on large State-supervised plantations. To stimulate agricultural output a 'Development Company' was set up in 1958. Its

activities include the construction of specially equipped packing stations for handling fruit and vegetables and despatching them swiftly to markets in Bordeaux and Toulouse.

The latter is an important route centre and the only other large industrial city in Aquitaine. Its flour-mills and textile, aircraft, engineering and chemical factories use hydro-electric power from the Ariège valley and natural gas from the fields at Lacq and St. Marcet. These gas fields are also linked by pipeline to Bordeaux and Nantes, and farther afield to Marseilles, Lyons and Paris. Petroleum from the Parentis field is carried by pipeline to Bordeaux for refining.

at Parentis and natural gas at Lacq and St. Marcet (*see map*). In 1955 an official 'Bordeaux and South-west Expansion Centre' was created to promote more effective use of the region's agricultural, commercial and industrial resources. Special encouragement is being given to industries using petroleum and natural gas as raw materials or sources of power.

> "If Aquitaine has been cut off from the two great industrial revolutions brought about by coal and hydro-electricity, it may today be said that, around Bordeaux, the entire Atlantic south-west coast is preparing to seize its chance with Lacq gas. The huge chemical and electro-metallurgical complex which is in process of being constructed on the gas field itself is, in fact, likely to give rise in the next few years to considerable industrial activity on the Pau-Bordeaux axis."*

* Jacques Chaban-Delmas, writing in *Progress*.

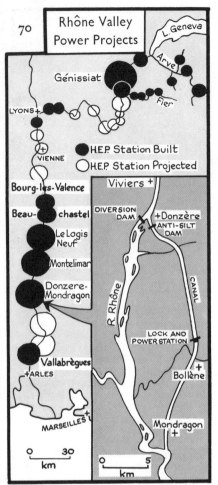

Rhône Valley Power Projects

Génissiat

LYONS

VIENNE

● H.E.P. Station Built
○ H.E.P. Station Projected

Bourg-lès-Valence

Beau-chastel

Le Logis Neuf

Montelimar

Donzere-Mondragon

Vallabrègues

+ARLES

MARSEILLES +

0   30
km

Viviers +

DIVERSION DAM

+Donzère
ANTI-SILT DAM

R. Rhône

CANAL

LOCK AND POWER STATION

Bollène +

Mondragon +

0   5
km

Major improvements to the Rhône-Doubs waterway will enable large barges to ply between the Rhine and Marseille by 1975.

THE RHÔNE-SAÔNE CORRIDOR has been, for well over 2000 years, a great natural routeway linking the Mediterranean coastlands with North-west Europe. Photo **A** shows the 'victory arch' erected in 49 B.C. by order of Julius Caesar at Orange, a day's march north of Avignon on the way to Northern Gaul. Past this arch now runs the main N7 motor road from Marseilles to Paris. Vital canals and railways (*p. 83*) also follow the Corridor, but today the Rhône Valley is becoming famous for quite different reasons. The river and its tributaries are being harnessed as part of a huge development plan to improve navigation and to provide water for power and irrigation.

Five of the largest projects have already been completed (*see map*). They include the great power scheme at Donzère-Mondragon: (*29*) describe this, referring to the map and to photo **B**. Electricity supplies from the power stations are attracting new industries. One example is the vast Rhône-Poulene chemical works 19 km down stream from Vienne. It is against this background of change that the way of life in this part of France must be studied.

The Position of Dijon

The appearance of the Corridor varies considerably from place to place. In the north the valley of the Sâone is broad and flat-floored. The river here is wide and comparatively placid, and provides a good waterway for heavy barges. A great number of these travel along the Sâone on the way to the Paris Basin *via* the canals shown on the map, right (*photo p. 82*). There is also a waterway to the Rhine *via* the River Doubs and a canal through the Belfort Gate. The climate of the Sâone valley is continental. (*Dijon: Jan.* −*1°c.; July 19°C.*; 625 mm *annual rainfall, over half falling in summer.*) Dairy farming is important, and typical crops include wheat, barley, sugar-beet and clover. The western edge of the valley is marked for part of its length by the Côte d'Or (*notes p. 62*), a limestone escarpment specially noted for the production of Burgundy wine. The vineyards on these elaborately terraced hillsides, with their sheltered position, sunny aspect and fertile soil, yield some of the finest wines in Europe. Dijon, the main market town of the Sâone valley, is also a very important route centre (*see map*) and has food-processing, chemical and metallurgical industries.

To the south of Lyon the Rhône valley consists of a series of farming basins linked by narrow, sometimes gorge-like sections. Each basin has its main market-town, many of which are old Roman settlements such as Valence and Avignon. South of Valence the climate becomes increasingly Mediterranean, and the summer drought makes irrigation essential. The cultivated land is mostly carved into tiny market-garden plots and intensively farmed with vines, wheat, tobacco, olives and various fruits and flowers. Windbreaks of cypress trees protect houses and crops from the *mistral*, a bitterly cold, dry wind which occasionally blows down the Rhône Valley with great force during winter and spring (*V:39*).

LYON

(*population 530 000*) is the regional capital for much of South-east France. Established as a fortress town by the Romans, it developed as an entrepôt for medieval trade and became Europe's main silk manufacturing centre in the 16th century. It flourished and expanded mainly because of its position as a major route focus. ((*30*) Show this by a sketch-map: include labelled arrows to mark each of the following routes:—*To Paris; To Rhine Valley; To Mediterranean ports; To Switzerland.*) Today about half of the workers in Lyons manufacture silk. Other major industries include synthetic textiles, metallurgy; electrical engineering, chemicals and foodstuffs processing.

'MEDITERRANEAN' FRANCE has a rather complicated relief. After looking at the map on the right one may ask what justification there can be for treating this part of France as a distinct geographical region. What is there in common, for example, between the rugged karst hills of the Garrigues, the dead-flat deltaic lowlands of the Rhône, the bay-head beaches of the Riviera and the steep foothills of the Maritime Alps? The answer is *climate*: all these districts have mild, wet winters and hot, dry summers with a near drought. (*31*) Draw climate charts for Montpellier and Nice. (*32*) For each place state (*a*) the highest and lowest mean monthly temperature; (*b*) the total mean annual rainfall and (*c*) the driest and wettest months. (*33*) Account for the cooler winters at Montpellier. (*Hint: Shelter.*)

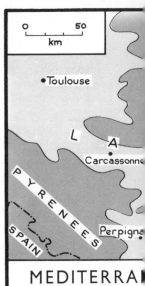

The blistering dry heat of the Mediterranean summer has given this part of France a distinctive landscape and way of life, so much so that the region has acquired a special name —the *Midi*. Traditionally the Midi is a land of irrigated peasant farming, of scrubby *maquis* vegetation (*p. 24*), of sun-drenched beaches and fierce autumn rainstorms. Today the Midi is also a land of rapid and

NICE	J	F	M	A	M	J	J	A	S	O	N	D
Temp. °C	8	9	11	14	17	21	23	23	20	16	12	8
Rainfall mm	63	55	68	55	60	40	15	23	60	145	110	78
MONTPELLIER												
Temp. °C.	5	7	9	13	16	20	23	22	19	14	9	6
Rainfall mm	78	68	60	58	70	45	23	50	75	103	85	60

Enjoying the sun—the beach at Cannes.

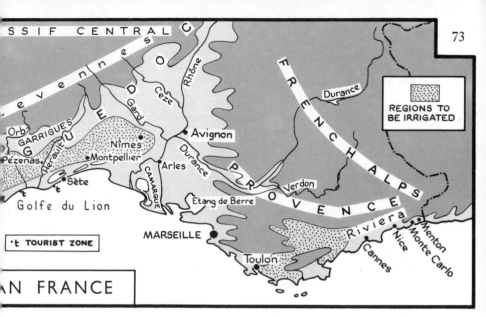

REGIONS TO BE IRRIGATED

·ţ TOURIST ZONE

…N FRANCE

far-reaching changes, notably in *Languedoc*. Here the land ascends in a series of gravel-covered terraces from a low-lying, lagoon-fringed coast to the steep slopes of the Massif Central. The coast, with its 160 kilometres of fine sandy beaches, is being developed to cater for 1 million tourists in six special zones (*see map*) separated by nature reserves. New motorways and link roads give easy access to recently built villas, hotels, camping sites and yacht marinas.

For centuries Languedoc has been famous for its wine. Indeed Hérault has two-thirds of its surface covered in vineyards and produces two-fifths of all French wine (mostly *vin ordinaire*). But many farms are too small for efficiency and much land suffers acutely from drought in summer. Furthermore, over-concentration on wine production has impoverished the soils and made the population too dependent on a single crop. However . . .

Using the sun—salt evaporation pans.

Fighting the sun—irrigation channels.

"Something is stirring in France's 'Deep South'. From the lower Rhône almost to the Spanish frontier, a vast stretch of sunbaked land is being transformed into a bountiful fruit and vegetable garden. ... Today an impressive network of dams, canals and irrigation ditches is reaching out across the region's 600,000-odd acres. By 1975, when the plan is complete, about 420,000 acres . . . will be under irrigation. The principal work is a 174´ wide canal carrying Rhône river water from a point north of Arles to Pézenas 100 miles away. . . . From this main waterway, more than 130 miles of tributary canals are being dug to link with farmers' pumps and sprinkler systems throughout the eastern half of the zone. For the west, the waters of the Orb and the Hérault are being harnessed. . . . Also under way . . . are . . . refrigerating centres, silos, slaughter houses, cattle-food factories, canning plants, markets, farm equipment and other ancillary industries. . . .

"An increasing flow of apples and pears is finding its way to Paris, the Common Market countries and London.... Soon loads of peaches, apricots, plums, melons, asparagus, cauliflowers and potatoes, as well as better-quality wines, will be rolling out of the south-west. An up-to-the-minute co-operative marketing complex at Nîmes is already showing what can be done, with teleprinters flashing hourly reports on Covent Garden, Milan, Brussels, Paris and Hambourg."*

A similar transformation is taking place in the *Camargue*, the 'island' formed by the two branches of the Rhône between Arles and the Mediterranean. Until recently this stony, swampy area, the remains of a salt lake, was best known as a bird sanctuary and the home of the wild bulls used in the Provence bull-rings. Today it produces most of the rice consumed annually in France.

"This barren and desolate region . . . is about 200,000 acres in extent. In the nineteenth century its inhabitants made some effort to drain off the salt water and irrigate their land, especially in the north, where they planted vines which produced wine of poor quality. . . . In 1946 2,500 acres were sown with rice; today the area is 57,000. Last year's production was worth nearly £4 million. . . . Five hundred miles of irrigation channels had to be dug, and pumping stations built along the Rhône and the principal channels. One of these pumping stations, near Arles, delivers 1,500 gallons of water a second. . . . The extent to which the Camargue has been transformed may be judged from the fact that consumption of electricity has increased fourfold in the past four years. The paddy is converted into saleable rice at three factories in the Arles neighbourhood."*

Provence, to the east of the Rhône, is more mountainous than Languedoc; the rocky, indented coast has bold promontories and

The climate, with its mild winters	?	allows a maximum of sunshine.
The south-facing aspect		make the region easily accessible.
The steep mountains to the north		makes the coast extraordinarily beautiful.
The combination of sandy bays, coves, mountains and azure sea		give protection from the cold 'Mistral'.
Modern road, rail and air services		attracts visitors throughout the year.

* The *Daily Telegraph*.

fine bay-head beaches. A mild climate and the development of fast modern transport have favoured the growth of two principal commercial activities:—(i) intensive cultivation of early spring vegetables and flowers; (ii) the tourist industry.

These two activities overlap, for well over 1 m. holiday-makers visit the Riviera every year, and the great resorts of Cannes, Nice, Monte Carlo and Menton are in themselves a valuable market. Vegetables, fruit and flowers are also sent to Paris and London. The tourist industry has grown enormously in the past 50 years.* (34) Make a list of the factors which have favoured its development on the Riviera by sorting out the 'Heads and Tails' opposite.

The different coastlines of Languedoc and Provence have influenced the growth and importance of French Mediterranean ports. Longshore drift carries mud and sand westwards from the Rhône delta, and over the centuries the coastal swamps and lagoons of Languedoc have gradually been silted up. Several ports which were important in Roman times have long since fallen into decay. Only Sète, with its fishing fleet and oil refinery, is now of any significance. East of the Rhône delta, away from danger of silting, lie the great commercial and industrial port of Marseilles and the naval base of Toulon.

*—and has now begun to affect CORSICA. This mountainous, sparsely-populated and under-developed island has no other important industry and no large towns; and only 2% of its area is under cultivation.

MARSEILLE (population 965 000) is the second city of France and her greatest commercial port, handling about one-quarter of all French maritime trade. Founded by the Greeks about 600 B.C., the port became really important only in the mid-19th century. The opening of the Suez Canal in 1869 gave Marseille easy access to Oriental markets, and the acquisition by France of a huge African empire brought in a prosperous colonial trade. Extension of the docks at first took place northwards, behind a great protective mole built parallel to the shore. Since 1918 new docks and industries have been located around the shores of the Étang de Berre, where vast oil refineries have recently been constructed. ((35) Where exactly? See map.) Marseille developed primarily because of her commanding position at the southern end of the Rhône-Saône Corridor: the port thus serves the Paris Basin, the Lyons industrial area and Switzerland. Major industries in Marseille include: *petroleum and vegetable oil refining, petro-chemicals, shipbuilding, soap and margarine production and steel manufacture.* A large crude steel plant at Fos has been built to give extra employment, to attract metallurgical industries and to develop steel exports to Southern Europe and North Africa. This 'tidewater' steelworks is entirely dependent on iron ores and coking coal brought in by sea (see also note, p. 187). Marseille is a major passenger port and her airport is the second busiest in France.

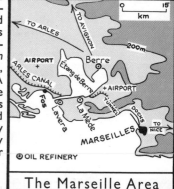

The Marseille Area

ARMORICA

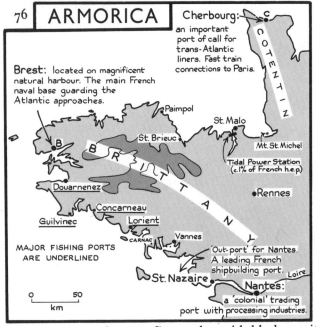

Cherbourg: an important port of call for trans-Atlantic liners. Fast train connections to Paris.

Brest: located on magnificent natural harbour. The main French naval base guarding the Atlantic approaches.

COTENTIN

B R I T T A N Y

Paimpol

St. Brieuc

St. Malo

Mt. St. Michel

Tidal Power Station (c. 1% of French h.e.p.)

Douarnenez

Rennes

Concarneau

Guilvinec

Lorient

CARNAC

Vannes

MAJOR FISHING PORTS ARE UNDERLINED

'Out-port' for Nantes. A leading French shipbuilding port.

St. Nazaire

Nantes

Loire

a 'colonial' trading port with processing industries.

0   50
km

ARMORICA is a Celtic word meaning 'the land of the sea', a very apt description of the Brittany and Cotentin peninsulas. In many respects this region is very different from the rest of the country. Geologically it consists partly of ridges of old, hard rocks ((*36*) *the remnants of which fold-mountain system?*), together with bleak granite uplands and many fertile valleys and secluded basins. For millions of years the exposed but resistant rocks of Armorica have been ceaselessly buffeted by Atlantic waves: the result is a magnificent rugged coastline. The harder rocks form bold cliffs and headlands, seawards of which for several kilometres run treacherous rocky shoals. (*37*) Explain why.

A slight rise in sea-level in recent geological times has caused the sea to flood the intervening river valleys, forming splendid rias. (*38*) Name five ports located on rias in Armorica.

The climate of Armorica is distinctly maritime. The prevailing Westerlies bring depressions and rain throughout the year. Many days are overcast and tourists are often disappointed to find the scenery shrouded in a fine drizzle. This disadvantage is partly offset by the equable temperatures. (*39*) Draw a climate chart for Brest and state (*a*) the total mean annual rainfall and (*b*) the mean annual range of temperature. (*40*) Compare and contrast,

Mont St. Michel—built on one of the granite islets that dot the rugged coast of Brittany.

BREST		J	F	M	A	M	J	J	A	S	O	N	D
Temp. °C		7	7	8	11	13	16	17	18	16	13	10	8
Rainfall mm		88	75	63	63	48	50	50	55	58	90	105	110

with explanations, the climate of Brest with that of Nice (*p. 72*).

The mild, moist climate favours dairy farming ((*41*) *why?*) and Armorica contains one in five of all French dairy cattle. One in five of all French pigs are also reared in Armorica—(*42*) suggest a reason. Fodder crops (mostly oats and barley) are grown, as well as rye and buckwheat for human consumption. Crops mature quickly on the sandstone soils of the coastal fringe and market-gardening is very important. Apple orchards are widespread and much cider is made. Every spring, huge quantities of small fruit and early vegetables (onions, potatoes, broccoli, etc.) are sent by fast trains to Paris and by boat from St. Malo to England. The southern margin of Brittany and the lower Loire Valley are especially productive. Farms are mostly small and the fields are protected from the frequent gales by tall hedges built of granite blocks and earth. The 'chequer-board' landscape which results is called *bocage*.

For centuries the people of Armorica have turned to the sea for an additional source of livelihood. *Lorient* is the second fishing port of France: other harbours with sizeable fishing fleets are underlined on the map. ((*43*) *Name them.*) Fish caught include mackerel, pollack, tunny and sardines, as well as shrimps, shellfish, lobsters and oysters. Some larger boats regularly visit the Grand Banks cod fisheries off Nova Scotia. Much of the interior of Armorica is barren and sparsely peopled, and all of the major towns except Rennes are located on or near the coast. Details of the more important of them are given on the map opposite.

Rennes (population 195 000) is the capital of Brittany and an important market town. It grew up at the centre of the largest and most fertile basin in Armorica. As a natural focus of many roads and rail routes it functions as a commercial and servicing centre for much of north-west France. In addition it has a famous university and a growing range of industries.

Brittany has a rich legacy of pre- and early historic monuments, such as these famous Megalithic stones at Carnac. For centuries the Celtic inhabitants of Brittany had more in common with the people of Cornwall, Wales and Ireland than with other Frenchmen. The Breton language, still widely spoken, closely resembles old Cornish.

HIGHLAND FRANCE

1. *THE MASSIF CENTRAL* is the most extensive region of high ground in France, covering one-sixth of the country. Its physical geography is indicated in the diagram and map: (*44*) study these, and then complete the following paragraph correctly:—

The Massif is $\frac{entirely}{mainly}$ composed of old, hard crystalline rocks such as g......, surrounded by a $\frac{broken}{complete}$ rim of Carboniferous rocks. The latter are mainly $\frac{limestones}{granites}$, but include some coal-bearing rocks, e.g. at Le C......, St. É...... and A..... In some places, notably in A......., lavas and other $\frac{volcanic}{sedimentary}$ rocks have forced their way to the surface through faults in the older formations. Most of the Massif lies between 1000 m and 2000 m, and its surface is for the most part $\frac{deeply\ indented}{gently\ undulating}$. The whole 'block' was tilted downwards to the $\frac{south-east}{north-west}$ by the earth-movements which built the Alps: hence $\frac{much}{all}$ of the higher ground lies in the Cevennes, which drop away dramatically by a f.... s.... to the plains of Languedoc. Many rivers cross the edges of the Massif by deep-cut gorges, especially in the limestone country of the C...... The Loire and its tributary the A..... are deeply entrenched in minor rift valleys.

Land-use in the Massif Central. As the Massif lies between three different climatic regions ((*45*) *name them*) its climate varies greatly from place to place. In summer the north-western districts have the characteristic warm, moist conditions of N.W. Europe, whilst the south and south-east are more typically 'Mediterranean'. Winters, however, except in the deeper valleys, are everywhere cold and snowfall is heavy. Precipitation exceeds 2000 mm on the higher ground.

The higher plateaus and ridges are largely given over to livestock farming: beef and dairy cattle on the damp pastures of the centre and north, sheep and goats on the scrubby karst uplands in the south. Roquefort cheese, made from ewes' milk, is a well-known product from the latter districts.

Arable farming varies according to altitude, soils and climate. On the high exposed plateaus of the north-west only rye, buckwheat and potatoes are important, but the fertile, sheltered valleys are planted with wheat. The rich, volcanic soils of the Loire and Allier valleys are the main farming areas. In the hotter south many valleys are terraced, and maize, vines, mulberries and olives are grown.

For many years younger people have left the Massif to seek better paid jobs elsewhere. To counter this depopulation new Regional Plans for Auvergne and Limousin aim to improve agriculture and forestry and to promote tourism and other industries. Notable developments include:— sheep and cattle breeding stations to raise the quality of local livestock; growing irrigated fodder crops in the Allier valley; upgrading highland pastures by drainage and by sowing better seeds; and the afforestation of 5000 ha of formerly unproductive hill country. Some hydro-electricity lakes are being developed as tourist centres, with holiday houses and villas and facilities for camping and caravanning.

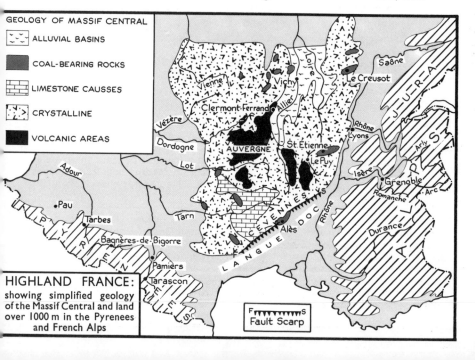

DIAGRAMMATIC SECTION ACROSS THE MASSIF CENTRAL

The coal basins of the Massif, although small, have given rise to several industrial districts. The most important of these lies around *St. Étienne*. The coal seams here are much broken by faulting, but the field has been vigorously exploited ever since the main Flanders coalfield was overrun during the 1914–18 War. About one-quarter of the 3·5 m. tonnes of coal produced annually is sent to the nearby Lyons industrial district. The rest is mostly used by factories in and around St. Étienne. Coke is produced for the local iron and steel works, which use pig iron from Lorraine to produce high-grade steels for armaments, vehicles, aircraft and other specialist engineering purposes. There are also chemical, silk and artificial-textile, glass and pottery works in the St. Étienne district. In addition St. Étienne is the market and shopping centre for the whole eastern margin of the Massif.

Le Creusot, the other main coalfield town, has a variety of engineering industries, the most notable products of which are armaments and locomotives. The *Alès* coalfield produces about 3 m. tons of coal annually. Industries there include high-grade steel making, chemicals and silk manufacture.

Clermont Ferrand is the second largest town in the Massif, and the main commercial and marketing centre of the prosperous Allier Valley. During the 19th century it became a very important rubber-tyre manufacturing town. Today it has the huge *Michelin* and *Bervougnan* rubber factories, and its other industries include engineering, chemicals, footwear and food-processing (especially jams and chocolate).

Power for the industrial towns of the Massif is obtained not only from thermal electricity generated on the coalfields but also from numerous h.e.p. stations in the valleys of the Lot, Tarn, Vézère and Dordogne. These stations account for about one-quarter of the total water-power produced in France. (*III: 187*).

The remoter parts of the Massif Central were somewhat cut off from the mainstream of French life, until this isolation was to a certain extent broken in recent years by the growth of tourism. Main tourist attractions include the remarkable volcanic 'necks' in the vicinity of Le Puy, the great limestone gorge of the upper Tarn and the warm mineral springs at Vichy, which is an inland 'spa' resort.

(*46*) Use the information in the last four paragraphs to draw a *large* labelled map, entitled 'The Massif Central: Farming, Industry and Main Towns'.

2. *THE FRENCH ALPS* form part of the great Alpine fold-mountain system described on page 10. As in all such glaciated mountain regions the highest and most rugged areas are uninhabitable ((*47*) *why?*), and the scant population is confined largely to the lower slopes and river valleys.

(*48*) Which two rivers and their tributaries drain virtually all of the French Alps?

About two-thirds of the hydro-electricity and one-third of the total power generated in France is produced in the Alps, including

---

FARMING IN THE FRENCH ALPS

Livestock rearing is the main agricultural pursuit throughout the French Alps, but the exact pattern of farming varies according to latitude. The main contrasts are summarised below. (*49*) Make a *large* copy of the map of the French Alps on page 79 and add the notes in their appropriate places.

Northern French Alps	Southern French Alps
'Continental' climate, with some snow-fields, glaciers and pine forests.	'Mediterranean' climate, with dry *karst* landscape.
DAIRY CATTLE on the green Alpine meadows, with mountain-valley *trans-humance*.	SHEEP for wool and skins/milk and cheese.
ARABLE FARMING in valleys; wheat, hay, potatoes, vines and deciduous fruit (*e.g. apples*).	ARABLE FARMING only on irrigated, terraced slopes; vines, wheat, maize, tobacco and sub-tropical fruit (*e.g. apricots*).

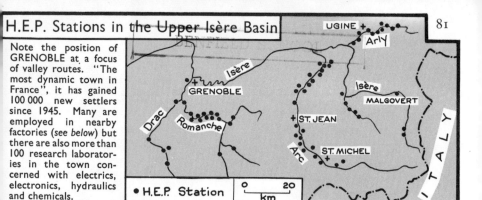

Note the position of GRENOBLE at a focus of valley routes. "The most dynamic town in France", it has gained 100 000 new settlers since 1945. Many are employed in nearby factories (*see below*) but there are also more than 100 research laboratories in the town concerned with electrics, electronics, hydraulics and chemicals.

the Rhône Valley (*map p. 70*). New schemes are constantly being implemented, notably in the drainage basin of the upper Isère. (*50*) How many power stations are located there already? The largest Alpine station is at Malgovert, opened in 1954, with an annual output of 520 million kWh. Most of the power produced by these stations is used by industries in the Alpine valleys, but some is fed into the 'grid' for distribution throughout France. The expanding Alpine industrial region has its focus and 'capital' at Grenoble. Long lines of factories, mostly concerned with electro-chemicals or electro-smelting, stretch along valleys such as the Isère, Arly, Romanche and middle Durance. Some traditional industries (e.g. textiles and glove- and watch-making) have been modernised, but the major works produce such goods as steel alloys, aluminium, carbide, explosives and fertilisers.

OTHER HIGHLAND REGIONS OF FRANCE

THE PYRENEES rise abruptly from the plain of South Aquitaine, presenting an almost continuous barrier of heavily glaciated folded ranges from the Bay of Biscay to the Mediterranean Sea. Although not so high as the Alps they are difficult to cross, especially in the centre where winter snowfall is heavy and all of the passes lie above 2000 m. (*51*) Describe from your atlas the paths followed by the rail routes linking France and Spain. Limestone is the predominant rock and thus karst scenery is commonplace, especially in the Eastern Pyrenees. ((*52*) Why there?)

On the high ground livestock farming, mostly sheep and goats, is the main source of income. Cattle are reared in the north-west, which is much wetter ((*53*) why?) and so has more luxuriant pastures. In the valleys, arable farming is possible on boulder clay and alluvial soils. Wheat, maize, vegetables, hay and vines are the main crops.

Here, too, many modern h.e.p. stations have been built: they produce about one-sixth of the entire French output of h.e.p. Cheap electricity has attracted electro-chemical and electro-metallurgical industries to the Pyrenean valleys. (*54*) Name the main industrial centres from the map on page 79.

THE JURA are a series of limestone plateaus and fold ranges, in places reaching over 1700 m. They have a substantial rainfall and so are well wooded, and there are extensive upland pastures.

Cattle rearing is the main source of livelihood, the animals being taken to upland pastures to graze between May and September, and brought back to sheltered valley farms for the winter. The region is well known for its dairy produce, notably Gruyère cheese. There are also traditional 'mountain' occupations such as woodcarving and making clocks and toys.

Barge traffic on the Rhône at Donzères. In France great use is made of inland waterways for moving bulky cargoes. Use the maps in this chapter and your atlas to draw a simple sketch map of the main canals of France.

GENERAL SUMMARY. A striking fact about France is that in spite of its very varied geography the country is politically well integrated. Power has been focused in Paris ever since France emerged as a nation-state, and the various provincial and city councils have far less control over day-to-day affairs than equivalent bodies in the United Kingdom. One important result of this centralisation is seen in the radial pattern of communications already noted (*p. 63*). (*55*) To what extent is the pattern of French railways (*see map*) influenced by relief? (*56*) How do you think the relief of the country has (*a*) favoured and (*b*) hindered the policy of centralisation?

The varied and complementary nature of its regions makes France much less dependent on foreign trade than some other economically advanced nations such as Great Britain and Italy. French industry depends on imports for a large proportion of the raw materials used, but the country is nearly self-sufficient in wheat and staple foodstuffs. The chief commodities bought from other countries are:—*foodstuffs* (19%); *fuel, oil, etc.* (17%); *machinery and transport equipment* (14%); *metals and metal goods* (11%); *textiles and textile raw materials* (9%); *chemicals* (5%); *others* (25%). The chief exports are:—*machinery and transport equipment* (33%); *metals and metal ores* (17%); *foodstuffs* (13%); *chemicals* (9%); *textiles* (7½%); *others* (20½%).

(*57*) Draw bar diagrams to illustrate this trade pattern.

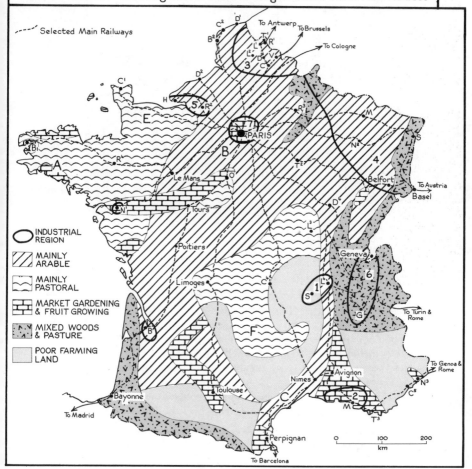

Selected Main Railways

INDUSTRIAL REGION

MAINLY ARABLE

MAINLY PASTORAL

MARKET GARDENING & FRUIT GROWING

MIXED WOODS & PASTURE

POOR FARMING LAND

This map shows the main farming regions of France. (58) Describe and explain what broad links you can see between it and the relief map (p. 59) and the rainfall map in your atlas. (59) Suggest, with reasons, the main types of crops you would expect to find growing at A, B and C. (60) What differences are there between the types of pastoral farming at E and F? (61) Account for the distribution of market gardening and fruit-growing regions.

About half of the 51 million inhabitants of France live in small market towns and villages and derive a livelihood from farming. An Englishman travelling through France is impressed by the peaceful rural scenes which still dominate the landscape; sprawling industrial conurbations —depressingly familiar in Britain—are few and far between.

The principal industrial regions of France are labelled 1–7. Identify each one from this list: *Rouen, Eastern, Lyons– St. Étienne, Nord (N.E. Coalfield), Alpine, Marseille, Paris.* Some important cities, isolated from the manufacturing regions, are indicated on the map by initials. (62) Identify each of them and name their principal industries/activities. (63) Of the main manufacturing regions, which obtain power primarily from (a) hydro-electricity? (b) local or nearby coalfields? (c) imported petroleum?

COLOGNE—1945

CHAPTER 4
West Germany

FOLLOWING the defeat of Nazi Germany in 1945 the boundaries of the former German Third Reich (*or 'Empire'*) were drastically revised. The main changes were:—

1. The loss of 114 000 square kilometres of territory to Poland and 15 000 square kilometres to the U.S.S.R. (*see map*). The lost territories included fertile agricultural lands in Pomerania and East Prussia, and the wealthy industrial region of the Upper Silesian coalfield.

2. The re-establishment of Austria (which had been absorbed into Nazi Germany in 1938) as an independent country.

3. The division of Germany into two political units:—(i) the

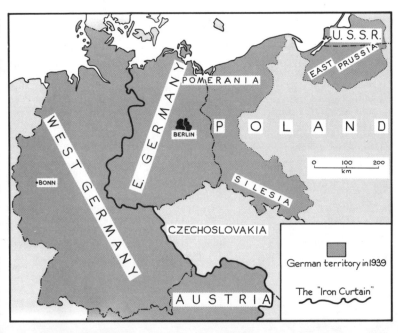

Federal German Republic, usually called West Germany, and (ii) the *German Democratic Republic* i.e. East Germany. Berlin, formerly the capital of all Germany, was itself divided into two parts; East Berlin became the capital of East Germany, and West Berlin, although entirely surrounded by East German territory, became part of the Federal Republic. The West German capital is at Bonn.

As East Germany is a Communist State and allied to the U.S.S.R., and West Germany is non-Communist and allied to the 'Western' Powers, the extraordinary situation of Berlin since 1945 has at times been a major source of international tension.

In 1945 much of Germany lay in ruins. Her main cities had been reduced to heaps of rubble by air bombardment and ground warfare. Roads, railways, bridges, port installations—all had suffered immense destruction. The dazed population lived on the verge of starvation and there were millions of homeless and unemployed. Yet today most Germans are both prosperous and well fed. The most spectacular recovery has taken place in West Germany, but conditions in East Germany also are steadily improving. How does one account for the remarkable transformation shown in the photographs above?

The answer lies partly in the traditional capacity of Germans for hard work, but the main reasons are found in the geography of the country. The maps on pages 11, 26 and 28 show that Germany contains some of the finest farmland in Europe, as well as major coalfields, mineral deposits and industrial centres. Economic recovery since 1945 has been based on the exploitation of these abundant natural resources. This has been done in fundamentally different ways on either side of the 'Iron Curtain', and so it is

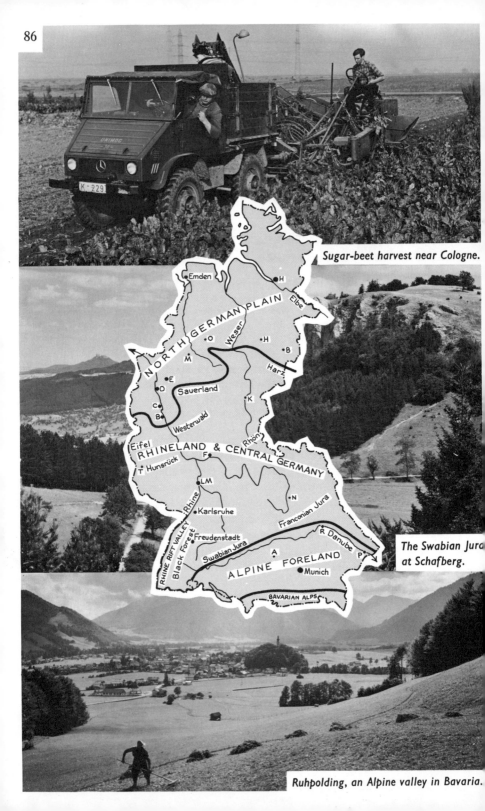

Sugar-beet harvest near Cologne.

NORTH GERMAN PLAIN

Emden

H

Elbe

Weser

O

M

H

B

Harz

E

D

Sauerland

K

C

B

Westerwald

Rhön

Eifel

RHINELAND & CENTRAL GERMANY

T Hunsrück

F

LM

N

Rhine

Karlsruhe

Franconian Jura

R. Danube

P

Black Forest

Freudenstadt

A

ALPINE FORELAND

RHINE RIFT VALLEY

Swabian Jura

Munich

BAVARIAN ALPS

The Swabian Jura
at Schafberg.

Ruhpolding, an Alpine valley in Bavaria.

EMDEN (Altitude 9m)	J	F	M	A	M	J	J	A	S	O	N	D
Temp. °C	1	2	3	7	11	15	17	16	14	10	4	2
Rainfall mm	53	40	50	40	50	60	80	85	60	73	60	63

KARLSRUHE (Altitude 133 m)	J	F	M	A	M	J	J	A	S	O	N	D
Temp. °C	1	2	5	10	14	17	19	18	15	10	5	1
Rainfall mm	48	43	50	58	55	70	75	78	73	63	60	63

convenient to describe the geography of West and East Germany separately. The rest of this chapter deals with West Germany; East Germany is described on pages 235–243.

The map opposite shows the main physical regions of West Germany:—(i) the North German Plain; (ii) the Highlands and Basins of the Rhineland and Central Germany; (iii) the Alpine Foreland and Bavarian Alps. The photographs illustrate some of the differences in structure and relief between these regions.

There are also noteworthy contrasts in their climates. The North Sea coastlands have a modified N.W. European climate, with rather cold winters. (*Compare the January mean temperature of Emden with that of Valentia on p. 15.*) Farther east, and inland, the climate becomes increasingly continental though conditions vary considerably according to altitude, aspect and shelter.

(*1*) Describe and explain the contrasts in (*a*) winter temperatures, (*b*) summer temperatures, (*c*) amount and seasonal distribution of rainfall between Emden and Karlsruhe. (*Re-read pp. 14–17 if necessary.*)

Karlsruhe is a sheltered town in the Rhine Rift Valley; Freudenstadt, only 72 kilometres away, lies in an exposed position in the mountains of the Black Forest. (*2*) Draw climate charts for both these places to illustrate differences in their climates.

(*3*) Compare the mean annual temperature range of Munich with that of Emden. Notice the distinctly continental aspects of Munich's climate—summer maximum of precipitation (nearly all convectional), cold winters and, in spite of the altitude, fairly high summer temperatures.

THE NORTH GERMAN PLAIN extends also into East Germany (*p. 235*) and forms part of the great European lowland stretching from the Paris Basin to the Russian steppes. On page 11 we saw that the German portion of this lowland was greatly affected by deposition during the Ice Ages. In West Germany a

FREUDENSTADT (Altitude 798 m)	J	F	M	A	M	J	J	A	S	O	N	D
Temp. °C	−1	−1	2	6	11	14	15	15	12	7	3	0
Rainfall mm	128	118	143	98	108	120	128	113	100	115	125	160

MUNICH (Altitude 580 m)	J	F	M	A	M	J	J	A	S	O	N	D
Temp. °C	−2	0	3	8	13	16	18	17	13	8	3	−1
Rainfall mm	40	33	48	70	93	120	118	100	93	58	43	45

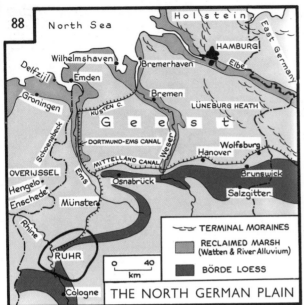

Reclaimed polder land near Emden. Note the ocean-going ship on the canal and the factories located beside the waterway.

belt of huge terminal moraines in Schleswig-Holstein marks the position of the edge of the Scandinavian ice-sheet at various stages during its final decay (*see map*). To the west and south-west lie extensive patches of boulder-clay, covered by glacial sands and gravels. These deposits form low, undulating plateaus called *geest*. The largest and highest area of *geest*, lying about 115 m above sea-level, is the Lüneburg Heath (*photo and text p. 25*). Cutting across the *geest* plateaus are the broad alluvial flood-plains of the Rivers Ems, Weser and Elbe. These plains, and the North Sea coastlands, are very low-lying and liable to flooding; photo **A** shows reclaimed polder land (*watten*) near Emden. The southern edge of the Plain, including the lowlands around Münster and Cologne, is covered by fertile loess.

Agriculture on the Plain varies greatly from place to place according to the nature and fertility of the soils. (4) Use the notes below to make a *large*, labelled map entitled 'The North German Plain—Farming and Land Use'.

GEEST DISTRICTS Many patches of heathland remain uncultivated. Some coniferous plantations and military training grounds, especially on Lüneburg Heath. Where heavily fertilised, the sandy soils yield good crops of CEREALS, POTATOES and SUGAR-BEET. On the lightest soils the main crops are rye and potatoes.	*RECLAIMED MARSH* of the coast and river estuaries. Wet clay and alluvial soils and wet, 'oceanic' climate favour rapid and lush growth of GRASS. Hence DAIRYING very important on both permanent and rotation PASTURE LAND. OATS and ROOT CROPS grown as animal fodder.	*LOESS BELT* of the 'Börde' —the best farmland in West Germany. INTENSIVE CULTIVATION on VERY FERTILE soils (*see map p. 11*). Main crops: WHEAT and SUGAR-BEET. Also many CATTLE, reared on pulp from sugar refineries and on FODDER CROPS grown in rotation.

Industry on the Plain is localised in two main groups of towns:—
(i) the inland market and commercial centres, and (ii) the ports.
The principal inland cities are *Brunswick, Hanover, Osnabrück* and
Münster. These are all acient towns which grew up originally as
trading foci for the rich agricultural lands of the Börde. Today
they have very important agricultural processing industries such
as flour-milling, sugar-refining, tanning and the making of starch,
glucose and industrial alcohol. ((5) *From what agricultural product in each case?*) Their industries also include light engineering,
farm implements, chemicals, synthetic rubber and vehicles.
Photo **B** shows the great Volkswagen car factory at Hanover.

Hamburg is the premier port of West Germany and a major
industrial centre. As early as the 17th century it had gained
control of traffic on the Elbe estuary. After the establishment of
the First German Empire in 1870 the growth of Hamburg was
spectacular—the population doubled between 1866 and 1890 to
over half a million: today it is 1 835 000. Several factors in
Hamburg's geographical position favoured its rapid growth:—

(*a*) Good transport via the Elbe waterway to a large hinterland in
eastern Germany and Bohemia (*map overleaf*).

(*b*) Access to the North Sea—since the Age of Discovery, a great
advantage over ports on the Baltic coast. (*Why?*)

(*c*) A small tidal range, making locks unnecessary and enabling
large ships to enter and leave the port at any time.

Before the Second World War Hamburg handled half of Germany's overseas trade. A great variety of cargoes was imported
for despatch by river and canal to Poland, Czechoslovakia, Austria
and Germany itself. Since 1945 it has lost much of its trade with

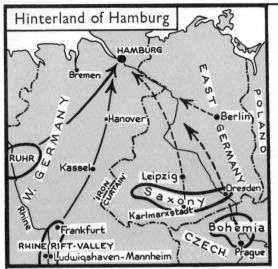

Hinterland of Hamburg

HAMBURG
Bremen
W. GERMANY
•Hanover
RUHR
Kassel•
Rhine
•Frankfurt
RHINE RIFT-VALLEY
••Ludwiqshaven-Mannheim
EAST GERMANY
POLAND
•Berlin
Leipzig•
IRON CURTAIN
S a x o n y
•Dresden
Karlmarxstadt
Bohemia
CZECH. Prague

Despite enormous war damage Hamburg was quickly rebuilt and regained its pre-war volume of trade by 1956. However, trade with territories which are now Communist fell by 53% and the port handles only 14% of the total trade of West Germany. Today over 60% of the port's traffic is connected with manufacturing in Greater Hamburg which has a working population, including commuters, of 1 120 000. Hamburg's industries have recently been greatly expanded and include chemicals, automobiles, aircraft, soap, tobacco and brewing. There are also very large oil refineries and shipbuilding yards. Hamburg is now the biggest West German industrial city after West Berlin. Recent or proposed new developments include:— a nuclear power station; an electro-steelworks; an aluminium rolling mill; a new canal link to the Mittelland Canal (map, p. 88); new autobahn connections to Hanover and Frankfurt; a container port at the mouth of the Elbe to handle modern bulk carriers.

eastern Europe, and faces keen competition from Bremen for West German trade.

Bremen (755 000), lying 64 kilòmetres upstream on the River Weser is also a major commercial and passenger port. Its trade and industries developed rapidly in the 19th century. Special links were forged with the Americas, for Bremen was the main port of embarkation for millions of Central European emigrants to the U.S.A. Returning ships brought in cargoes of tobacco and cotton, thus forming the basis of (still flourishing) tobacco and textile industries. Today Bremen is the major cotton importing centre for Central Europe, and also has engineering and shipbuilding yards. Since 1930 much trade has been channelled through the specially constructed outport of Bremerhaven, for the approach upstream to the old port is too shallow for large vessels. Trade with the Ruhr is conducted by rail, road and via the canal link to the Ems shown on the map on page 88.

Of the lesser West German ports *Emden* imports large quantities of iron ore and timber for despatch by canal to the Ruhr; *Bremerhaven* is one of Europe's greatest fishing ports, and *Wilhemshaven* is being developed as an oil terminal, with pipelines to the Ruhr.

West Berlin, with 2·2 million inhabitants and a working population of 300 000 is West Germany's greatest industrial city. In

The 'Iron Curtain'—Potsdamer Platz, Berlin.

1945, when Berlin was divided into its western and eastern segments, West Berlin was fortunate to contain the greater share of industrial plant. In spite of enormous wartime devastation these industries have been revived and considerably expanded. Even so, the beleaguered city has faced grave unemployment problems —thousands of workers in government offices lost their jobs when Berlin ceased to be the West German capital, and since 1949 refugees from East Germany have swollen the city's resident population by 200 000. New industrial housing estates have lessened the employment and housing difficulties. The city's principal industries are listed below together with the percentage of West Berlin's total industrial output for which they are responsible. (6) Draw bar diagrams to illustrate these figures.

Because of its special political status West Berlin is almost entirely cut off from its natural hinterland. Today 98% of West Berlin's exports go to West Germany, which in turn supplies the city with virtually all its foodstuffs and industrial raw materials. This trade goes by road (41%), rail (27%), canal (31%) and air (1%) along a few strictly controlled routes through East Germany. West Berlin's imports considerably exceed her exports, and so the city relies heavily on financial aid from the West German Government. The precarious economic position of West Berlin was demonstrated during the blockade and famous air-lift of 1948–9, when more than 2 million tonnes of commodities (mostly coal and food) were sent into the city by air from West Germany. Although the political division of Germany and of Berlin has never been ratified by international treaty, it seems unlikely that any change will occur in the near future.

Main Industries of WEST BERLIN	
Electrical equipment	(29%)
Foodstuffs and luxury goods	(18%)
Clothing	(12%)
Machinery	(10%)
Chemicals	(7%)
Miscellaneous	(24%)

91

(i) *THE NORTHERN RHINELAND* is a region of ancient sandstone and slate plateaus, deeply dissected into blocks by the Rhine and its tributary streams. From Bingen to Bonn the Rhine slices through the uplands by a spectacular gorge, part of which is shown in the photograph. Despite its forbidding appearance—in places the walls rise sheer from the river for hundreds of metres—the Rhine Gorge has been a focus of human activities for centuries. In medieval times it lay on a vital trade route linking the North Italian ports to Flanders. Valuable cargoes passing downstream were plundered or taxed by robber barons who lived in the 'fairy-tale' castles, perched on impregnable crags, which still line its banks. Today traffic in the Gorge is greater than ever, for through it pours most of the trade of the Rift Valley industrial towns and Switzerland (*p. 134*). Day and night the river is churned by an endless succession of shipping, including barge 'trains' towed by tugs. Write a full description of the scene shown in the photo. Notice that in spite of obvious difficulties main roads have been constructed on embankments at the foot of each valley wall. Main railways also follow the Gorge on both sides, cutting through the more difficult sections by tunnels.

The Moselle Gorge (*see map*) is also becoming an important artery of trade. An ambitious scheme to canalize the Moselle, opening it up to heavy barge traffic along the 176 kilometre stretch between the Saar and the Rhine was completed in 1965. The improved waterway is able

"to accommodate massive diesel-driven barges carrying iron ore from Lorraine to the Ruhr steel mills, and oil, timber and coal between France and Germany. . . . Simultaneously with the canalisation

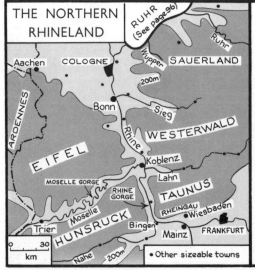

THE NORTHERN RHINELAND

Aachen COLOGNE · RUHR (See page 96) Ruhr SAUERLAND Wupper 200m Bonn Sieg WESTERWALD Rhine Koblenz Lahn EIFEL ARDENNES MOSELLE GORGE RHINE GORGE TAUNUS Moselle RHEINGAU Wiesbaden HUNSRUCK Bingen Trier Mainz FRANKFURT Nahe 200m

0 30
km

• Other sizeable towns

The largest town in the Rhine Gorge is Koblenz. (*13*) Suggest how its position favoured its growth and development as a market centre and river port. (*14*) What do you notice about the position of the other main towns of the Rhine Highlands? (*15*) What does this suggest about (*a*) the attractiveness of the uplands for settlement and (*b*) the importance of the Rhine as a routeway?

Bonn is a university and cathedral city whose importance has greatly increased since it became the capital of West Germany. The major industrial city of Cologne is referred to overleaf.

project a modern drainage system was laid throughout the valley and turbines installed at each weir now provide hydro-electric power for the villages. The scheme has brought not only modern amenities but increased trade, as the big Rhine passenger steamers now have access to the Moselle. In fact tourists are able to contemplate the marvels of modern engineering in a landscape moulded by the age-old technique of vine culture." (After Mary Howarth, writing in the *Daily Telegraph*.)

Wines from the Moselle valley, the Rhine Gorge and the Rheingau (*see map*) are indeed famous. In all these districts the south-facing slopes, even those which are near-precipitous, are terraced for vineyards. Can you spot any in the photograph? Cultivation of such land is a slow and laborious business: soil, manure and insecticides have to be carried up, and the harvested grapes brought down, along steep, narrow paths and steps. The plateau surfaces lie mostly above 500 m; cultivation there is hindered by exposure to wind, heavy rain, low summer temperatures and thin, infertile soils. Large areas, especially in the Westerwald and Taunus, remain forested. Where the land has been cleared sheep and cattle are reared on poor upland pastures. Only in the more favoured localities, notably in the Eifel where soils are comparatively fertile, is arable farming possible. Crops include rye, oats, barley and some wheat. At its northern end the Rhine Gorge opens into the Cologne 'bay', an intensively cultivated loess lowland (*p. 88*) with fields of wheat, sugar-beet and potatoes, together with pastures and orchards.

The Rhine Gorge at the Lorelei, near Bingen

THE RHINELAND and CENTRAL GERMANY

(ii) *THE SOUTHERN RHINELAND.* We saw on page 87 that summers in the Rift Valley are warmer than in most parts of Germany. The diagram on page 67 shows that some of the river terraces and lower flanks of the uplands also have a covering of fertile loess. *Farming* is intensive and the countryside is closely settled. Crops include maize, vines, tobacco and many varieties of fruit. Cultivation also extends up the valleys of the Neckar and Main. Market gardening is widespread, particularly in the area west of Heidelberg. The contrast between cultivated valley and forested upland is typical of the Southern Rhineland. (7) Make a sketch of the photograph opposite, adding the following labels in their appropriate places; pasture (cattle and sheep), arable patches in valley, forested hills.

The most extensive areas of *forest* occur in the Schwarzwald (Black Forest) and Odenwald. Although the soils of these largely granite plateaus are characteristically infertile, the heavy rainfall (*chart p. 87*) encourages tree growth and most of the rounded summits have a dense covering of pines. The Schwarzwald is one of Germany's main sources of timber. There is also a centuries-old wood-carving industry; village craftsmen produce toys, clocks and souvenirs for sale to the many tourists who visit this picturesque region. Freiburg is the main tourist centre.

As well as its flourishing farming and forestry, the Southern Rhineland contains several important towns. From the map, (8) suggest a political reason why few towns stand on the banks of the Rhine between Basel and Lauterbourg and (9) state what advantage of location favoured the development of Mainz and Mannheim-Ludwigshafen as river ports.

MAIN TOWNS OF THE SOUTHERN RHINELAND

Mannheim-Ludwigshafen is a major river-port, with 48 km of wharves. The twin towns (facing each other across the Rhine) form the greatest German inland harbour after Duisburg, and the largest commercial and industrial conurbation in S.W. Germany. The major industry is petro-chemicals, but there are also steel, electrical engineering and rubber works. Raw material such as coal and tar can be easily and cheaply assembled by river.

Frankfurt-am-Main is an ancient cathedral city and a focus of several very important rail, road and river routes (*see map*). It is a major river port and since the Middle Ages, when it became the site of great annual trade fairs, has been an industrial and commercial centre. Since the 17th century Frankfurt has been famous as a banking centre. Its varied industries include electrical engineering, machinery, chemicals, cars and oil-refining.

Mainz is an important communications centre with large-scale industries including engineering, chemicals and cement.

Karlsruhe was founded by the Prince of Baden in 1715. Today it is a river port with textile, engineering and electrical industries.

Wiesbaden is a tourist town and one of the main centres of the Rhineland wine industry.

Heidelberg is a picturesque tourist town with a world-famous university.

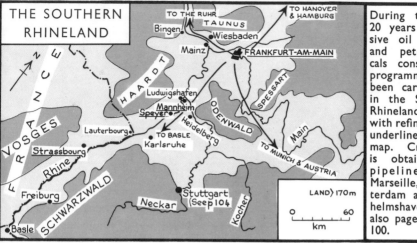

THE SOUTHERN RHINELAND

TO THE RUHR
TO HANOVER & HAMBURG
TAUNUS
Bingen
Wiesbaden
Mainz
FRANKFURT-AM-MAIN
HAARDT
Ludwigshafen
SPESSART
Mannheim
Speyer
ODENWALD
Lauterbourg
Heidelberg
TO BASLE
Karlsruhe
Main
Strassbourg
TO MUNICH & AUSTRIA
Rhine
VOSGES
FRANCE
Freiburg
SCHWARZWALD
Neckar
Stuttgart
(See p 104)
Kocher
Basle

LAND〉170m

0 ————— 60
km

During the past 20 years a massive oil refinery and petrochemicals construction programme has been carried out in the Southern Rhineland. Towns with refineries are underlined on the map. Crude oil is obtained by pipeline from Marseille, Rotterdam and Wilhelmshaven. See also pages 28 and 100.

THE SOUTHERN RHINELAND consists of the Rhine Rift Valley and its adjacent uplands. The latter are blocks (**horsts**) which were forced up on each side of the parallel lines of faults marking the edge of the Rift Valley. (*10*) Draw a diagram to illustrate this, referring to the sketch on page 67. The Rift Valley extends from Basel to Bingen. (*11*) Work out from the map its approximate length and breadth. The floor of the Valley is nearly flat, and formerly the River Rhine frequently burst its banks, inundating the broad flood-plain. Now the river has been straightened and confined between embankments, and serious floods are uncommon. (*12*) Suggest two reasons why these measures have also improved navigation.

The Schwarzwald near Freiburg.

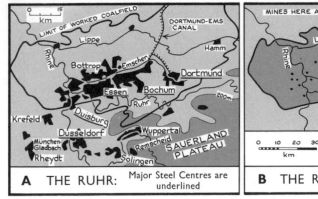

A   THE RUHR: Major Steel Centres are underlined

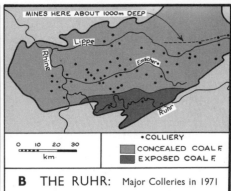

B   THE RUHR: Major Collieries in 1971

(iii) *THE RUHR BASIN* contains the greatest coalfield and the most concentrated industrial area in the whole of Europe. It takes its name from a small tributary of the Rhine, but industry has long since spread from the Ruhr valley to sprawl over an area of some 3900 square kilometres. The varied industrial activities of the Ruhr, upon which West Germany's prosperity chiefly depends, are based on coal, iron and steel. These two pages deal with the coal and steel industries: study them carefully and then (*16*) complete the following passage:—

The Ruhr coalfield lies along the northern flanks of the...... Plateau. The coal measures are at ground level in the valley of the ...... and dip gently $\frac{northwards}{southwards}$ beneath an 'overburden' of s......, c...... and glacial deposits. Hence the depth of coal-mines $\frac{decreases}{increases}$ steadily towards the north. The earliest coal workings in this region were in the ...... valley, but as these became exhausted mining spread into the $\frac{concealed}{exposed}$ coalfield. The northward drift of mining in the past century is indicated by the fact that in 1860 there were only .... large collieries on the concealed coalfield, whereas today there are .... Also the most northerly pits are now ...... kilometres farther within the concealed field, and ...... metres deeper, than in 1860. The great expansion in mining during this period is shown by the ....% increase in coal production. Coking coal makes up ....% of this output, and the presence of this very valuable product favoured the growth of an immense iron and steel industry. The early Ruhr iron and steel mills used local 'black-band' ores from the coal measures. Today over ......% of the ore used there is imported, the main foreign supplier being

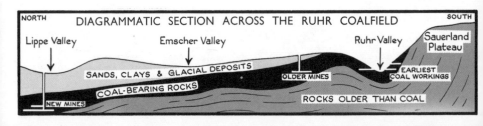

DIAGRAMMATIC SECTION ACROSS THE RUHR COALFIELD

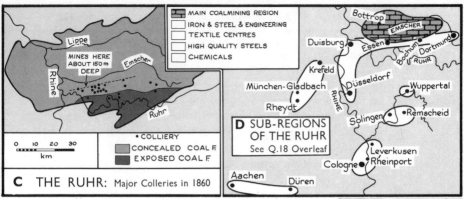

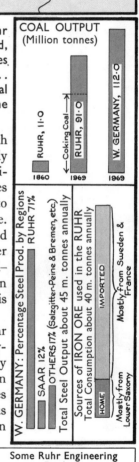

S,....... Much scrap metal is also used in the Ruhr blast furnaces and some home-produced ore is supplied, ...astly from L .... S....... The main steel centres are E......, D......, D......, D...... and B....... The Ruhr supplies no less than ......% of the total West German output of coal and ......% of the country's crude steel.

The development of the Ruhr steel industry, with its vital dependence on imported ore, was also greatly helped by extraordinarily favourable transport facilities. The Rhine is navigable for the largest barges and even for small ocean-going vessels right up to Duisburg, which is the largest inland port in Europe. A complex network of inland waterways, railways and roads also converges on the Ruhr from all over western Europe. Notice in particular the Dortmund–Ems Canal, which was built to give an all-German waterway for Ruhr trade; much Swedish iron ore is imported via this route.

Engineering is widespread throughout the Ruhr region, even in the villages of the adjoining countryside. The raw materials of this industry are chiefly varieties of steel, and so it is natural that the main steel-making towns in the Ruhr are also major centres of heavy and light engineering. Other famous centres are Remscheid and Solingen, the latter town sometimes being called 'The Sheffield of Germany'. Both these towns had a small-scale steel industry, using ore from the Rhine Highlands, centuries before the industrialisation of the Ruhr. Today they specialise in the production

97

Some Ruhr Engineering Products	
Heavy	*Light*
Bridges	Machine-tools
Boilers	Cutlery
Winding gear	Computers
Structural gear	Hardware
Agricultural machinery	Automobiles
Locomotives	Small-arms

Main Cotton Towns*	Main Woollen Towns	Silk and Synthetics
Wuppertal Krefeld München-Gladbach Rheydt	Aachen Wuppertal Düren Krefeld	München-Gladbach Rheydt Krefeld Wuppertal Dusseldorf Cologne

* Not strictly speaking in the Ruhr, but located just to the north, are the important cotton manufacturing towns of Bocholt, Gronau and Nordhorn.

of high-quality steels for such goods as cutlery, machine tools and surgical instruments. The tremendous variety of engineering products of the Ruhr region is indicated by the list on page 97.

Textiles have been made in the Rhine–Ruhr area for centuries. In the days of direct water-power the fast-flowing streams of the centres of production are indicated in the table above:— (*17*) show these notes in map form.

Chemicals are manufactured in many Ruhr towns, but the main centres are in and around Cologne, with vast works at Leverkusen and Rheinport (*Map D, p. 97 & III: 229*). The principal raw materials are coke-oven by-products, brought upstream from the Ruhr, and brown coal (lignite) obtained from open-cast pits to the south-west of Cologne. A huge petro-chemical industry has recently developed at Rheinport, using petroleum brought by pipelines from Rotterdam and Wilhemshaven (*see map*).

To meet an ever-increasing demand for petroleum products, new oil refineries have also been built in the western Ruhr (*see map*). Table Y shows how rapidly the refining capacity of the Ruhr region has expanded during recent years. At the same time the demand for coal has declined abruptly (*see Table p. 100*). These trends in the coal and petroleum industries form a critical problem for the Germans.

Since 1958 it has become increasingly obvious that German coal cannot compete with its rivals—oil, natural gas, imports of cheap American coal, and nuclear energy. The great age of Ruhr coal is apparently over, and a major transformation of the region's economy is in progress. The coal industry is being contracted and modernized: in 1957, for example, 400 000 Ruhr coalminers produced 149 million tonnes of coal, whereas in 1967 an output of 112 million tonnes was achieved with a work force of 190 000 men. Even so a further 80 000 miners face redundancy, and mines are now closed or are closing all over the Ruhr.

Bochum, for example, was until recently a typical Ruhr town, dependent on its coalmining and heavy steel industries. In 1952 more than one-half of Bochum's workmen were employed in 21 local coal pits: today only one pit remains open and the

EXAMPLES OF NEW & EXPANDING INDUSTRIES IN THE RUHR

> Oil refining and petrochemicals, e.g. plastics, synthetic fibres, ethylene (*Cologne*)
> Heavy engineering, e.g. steel tubes (*Dusseldorf*)
> Precision engineering and electronics (*Mulheim*)
> Nuclear engineering (*Essen*)
> Synthetic rubber (*Marl*)

99

colliery labour force has shrunk from 77 000 to 4500. Fifteen thousand Bochum miners became redundant in two years and at the same time many workers in the town's steelworks lost their jobs due to the introduction of labour-saving techniques and modern machinery. Unemployment has been avoided partly by retraining schemes whereby ex-colliery workers acquire new skills, and also by the location in Bochum of a bewildering range of new 'light' engineering industries, many of them deliberately sited on abandoned mine

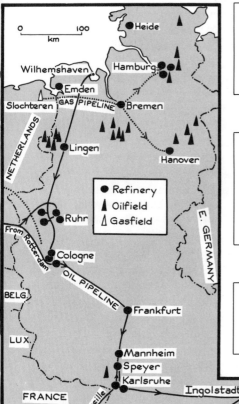

W. Germany: Oilfields, Pipelines and Oil Refineries

TABLE X
WEST GERMAN OIL PRODUCTION
(Million tonnes)

1946	1·1	1965	7·8
1950	1·1	1968	8·0
1958	4·4	1970	c. 8·4
1962	6·8		

(*18*) Use these figures to make a chart. Project the chart to give an estimate of the likely production in 1980.

TABLE Y
RUHR: OIL REFINING CAPACITY
(million tonnes)

1956	1960	1968	1971
6·3	23·8	38·6	c. 43·0

Continuous expansion of Ruhr oil-refining capacity is predicted up to 1980. The total value of products from these refineries already equals that of the Ruhr bituminous coal output, but the highly automated oil industry provides jobs for only a tiny fraction of those displaced from the closing collieries.

TABLE Z
PERCENTAGE REFINING CAPACITY
IN WEST GERMANY

Area	1951	1970
Coast	46	18
Lower Saxony	12	7
Rhine-Ruhr	41	38
South and Southwest	1	37

W. German inland refineries obtain most of their oil by pipeline from (which?) ports on both the North and Mediterranean Seas. Note the concentration of new refineries in Bavaria.

WEST GERMANY: TOTAL PRIMARY ENERGY CONSUMPTION			
	Coal	Oil	Total in coal equivalent
1950	72·5%	5·1%	126m. tonnes
1965	42·4%	41·4%	271m. tonnes
1970 (estimated)	35·0%	52·0%	300m. tonnes

workings. Foremost of the new plants are two large vehicle assembly works (*OpelKadett*), which employ 17 500 people. i.e. *four times* the number in the single remaining colliery. *

Similar changes are taking place throughout the Ruhr, and new industries such as those indicated above are growing very fast. Generous government subsidies are made available to help firms to locate new plants within the Ruhr, as well as to clear building sites, expand public utilities such as gas, water and electricity and increase bus and rail transport for commuters. The new 'growth' industries require highly-skilled technicians and managers, and so government money is also being spent on schools, technical colleges and universities. (*19*) Make bar diagrams to illustrate the table below, and comment on the relative importance today of coalmining and the steel industry.

This account of the Ruhr shows that sub-regions within the great conurbation tend to specialise in particular industries. The extent of these sub-regions is indicated on Map **D** (*p. 97*). (*20*) Make a *large* copy of this map, use different colours for the key and shade the remaining sub-regions appropriately.

NORTH RHINE WESTPHALIA (i.e. mostly the Ruhr), PERCENTAGE DISTRIBUTION OF THE WORKING POPULATION BY INDUSTRIES

Miscellaneous Metal Manufacture	17·2
Engineering	12·4
Textiles and Clothing	10·2
Coal Mining	10·2
Chemicals	8·2
Iron and Steel	8·1
Electrical Engineering	7·0
Motor Vehicles	3·3
Petro-chemicals	0·3
Others	22·7

Industrial developments similar to those of the Ruhr, and likewise underpinned by a massive increase in petroleum refining, are taking place in the Southern Rhineland ('*Rhine-Main*' map, *p. 95*). Here the petro-chemical and engineering industries (*see opposite, top*) are expanding so quickly that Frankfurt has become the main financial and foreign business centre of West Germany. Frankfurt-Wiesbaden is in fact a snowballing conurbation of over 2 million people, on which business men converge from all over the world. Frankfurt airport is now the largest in Europe, handling more than 7 million passengers a year (21 million are predicted for 1977).

Petro-chemicals—especially synthetics, industrial fibres and pharmaceutical goods. New chemical works at Hoechst give employment for 23 000 people. Chemicals make up 42% of the exports of the Frankfurt area, and this city contains many chemical and pharmaceutical research institutions.

Engineering—especially printing and paper-making machinery, pipes, hoists, cranes, pumps and machine tools.

Banking and Finance—Rhine-Main cities handle over a third of West German banking and about half of the country's foreign exchange. No fewer than 75 foreign banks have premises in Frankfurt.

Food-Processing—especially the manufacture of dairy products, sugar, spirits and tobacco.

The map below shows the location of two other coalfields of West Germany. *The Saarland* is the name of a portion of the basin of the River Saar, a south-bank tributary of the Moselle. It contains a small but valuable coalfield which became the basis of a prosperous iron and steel industry in the late 19th century. From 1919 until 1957 the Saarland was a bone of contention between France and Germany, each of which administered the territory according to the fortunes of war. In 1957 a settlement was reached whereby the Saarland, whose inhabitants are overwhelmingly German-speaking, became part of West Germany, although the French are allowed to work the collieries near the frontier until 1980. The main coal-mines lie in four small valleys converging on Saarbrücken, which is the largest town on the coalfield and the main business and administrative centre. The principal iron and steel works are in Saarbrücken and Neunkirchen. Most of the iron ore used there is brought from Lorraine, and local supplies of coking coal are supplemented by imports from the Ruhr.

Aachen lies to the south of a small coalfield, a westward extension of that in the Ruhr. The ancient city has old-established zinc-smelting and woollen industries (originally using raw materials from the nearby Ardennes). Using imported wool, its textile industry greatly expanded as power became available from the nearby coalfield. Eight collieries near the city now produce some 12 million tonnes of hard (non-coking) coal annually. A new refinery in Saarbrucken is supplied with oil by pipeline from Strasbourg.

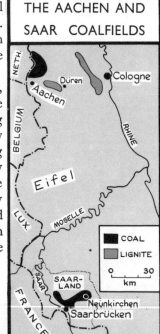

THE AACHEN AND SAAR COALFIELDS

COAL
LIGNITE

0   30
km

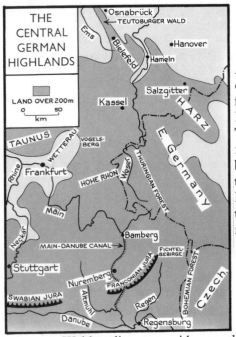

THE CENTRAL GERMAN HIGHLANDS

LAND OVER 200m
0   50
km

(iv) *THE CENTRAL HIGH-LANDS* of West Germany form a complex relief region stretching from the loess belt just south of Hanover to the Danube Valley. They lie mainly between 300 m and 1000 m and are deeply dissected by rivers draining in three directions:—to the North Sea; to the Rhine; to the Danube. (*21*) From the map state which of these is reached by following rivers:—*Ems, Weser, Regen, Neckar, Altmühl.*

The landscape varies considerably from place to place according to the composition and structure of the underlying rocks. Below we see part of the Teutoburger Wald, a limestone ridge south-east of Osnabrück. This is the most prominent of a series of chalk, sandstone and limestone ridges, separated by clay vales, which makes the Bielefeld-Hameln-Salzgitter district reminiscent of S.E. England. The Vogelsberg and the Hohe-Rhön (*see map*) are both volcanic in origin. The former is a basalt plateau with typically level sky-lines and precipitous valley slopes: the latter is the worn-down remnant of a long extinct volcano. The eastern margin of the regions is marked by a series of block mountains, such as the Bohemian Forest, Fichtelgebirge and Harz (*see page 7*). To the south lies another series of scarplands, with two magnificent escarp-

ments in the Swabian and Franconian Jura. The photo below shows part of the latter near Eschendorf. Notice that the scarp here is covered in vineyards. (*22*) Can you infer the direction in which the photographer was facing?

Throughout this region the uplands are largely forested with both coniferous and broad-leaved trees such as pine, spruce, larch, oak and beech. The valuable timber gave rise here to ancient wood-carving trades as well as modern industries producing constructional woods, pit-props, pulp, paper and fuel-wood. Forestry is highly developed and organised in West Germany, and timber from the Central Highlands supplies 25% of home demand. The valleys and basins within the uplands are fertile and well cultivated, and the landscape is dotted with prosperous farm communities. A great variety of crops is grown including wheat, oats, rye, potatoes and sugar-beet. In especially favoured spots, such as south-facing slopes and sheltered hollows, there are vineyards, orchards and fields of hops and tobacco. Cattle and sheep from the valley farms are taken in summer to browse on the pastures of lower hill slopes.

The principal towns in the Central Highlands originated as market centres. They are located in the centres of basins or at bridging places on the rivers, i.e. at the natural foci of communications. In medieval times, when the German lands consisted of many small principalities and kingdoms, several of them became state-capitals, e.g. Kassel was the capital of Hesse-Nassau, and Stuttgart the capital of Württemberg. These Central German towns, like those of the Rhine valley and Bavaria, benefited from their position astride medieval trade routes. Details of their growth and present importance are given overleaf.

Kassel is a route focus at the northern end of the Wetterau, an historic corridor ((23) *between which uplands?*) linking the Rhine Rift Valley with the North German Plain. Road and rail routes converge on the city from the Ruhr; from Hamburg and Bremen via Hanover; from Brunswick; from the Rhine Rift Valley; and from Bavaria. These routes are indicated by arrows on the map below. (24) Make a *large* copy of the map and add appropriate labels to the arrows. Kassel is an important market, commercial and administrative centre and has agricultural engineering industries.

Stuttgart is the main market and administrative centre of the fertile Neckar Valley. In medieval times it prospered by virtue of its position on trade routes leading from Northern Italy and the Danube Valley to the Rhine Valley. It is still a thriving city, with a great range of industries including chemicals, textiles, and both heavy and light engineering. In Stuttgart, as in other cities of Central and South Germany, the traditional skills and painstaking craftsmanship associated with the wood-carving and clock-making trades are reflected in the factory products. These include fine metalwork, musical instruments, ball-bearings and leather goods.

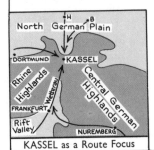

KASSEL as a Route Focus

Nuremberg (Nürnberg) lies in a natural route focus and on the Main–Danube Canal (*map p. 102*). It was a great medieval trading city and today has important heavy and light engineering manufactures including diesel engines, cranes and motor cycles.

THE ALPINE FORELAND AND BAVARIAN ALPS

are described and illustrated in the diagram below. (*Photo p. 86.*) (23) Use this information to write a geographical account of a journey direct from Ingolstadt to Oberammergau. In your description pay particular attention to the changes in the geology, scenery, agriculture and other activities in the districts through which you would pass.

South Germany is for the most part thinly populated, but there are several large and important cities, the location of which is shown on the map. These cities owe their origin and growth to their position on two ancient routeways, (i) from Northern

Danube Valley
Marsh and river-pasture on alluvial bottom lands. Arable farming on loess-covered river terraces. Ancient bridge-towns and market centres, e.g. Ulm, Ingolstadt and Regensburg, at foci of communications.

Bavarian Plateau
An undulating sandstone plateau, largely covered by glacial sands and clays brought from the south by ice-sheets during the Great Ice Age. Gentle northward gradient from 1000 m in Alpine foothills to 300 m in Danube Valley. Poor drainage—many peat-bogs (*moosen*) and lakes. Agriculture rather poor because of elevation, inferior soils, cool, moist climate and northerly aspect. Much permanent grassland in the south. Extensive coniferous forests, especially on the sandy moraines. Some arable farming on loess districts in the north. Main crops: oats, rye, potatoes, grass. Predominantly a region of peasant farmers.

GROUND MORAINE

LOESS & ALLUVIUM

DIAGRAMMATIC SECTION FROM THE DANUBE VALLEY

Munich (München), with a population of one million, is by far the largest city in South Germany, and the third largest in the country. It grew up on a gravel terrace beside the River Isar, surrounded by forests and infertile heath. Although it lay on important trans-Alpine routes (see map) its trade was for centuries overshadowed by that of Augsburg. When in the early 19th century it was made the capital of Bavaria, Munich acquired considerable importance as an administrative and cultural centre. More recently its commerce and industries greatly expanded following the construction of railways (especially those from Northern Italy via the Brenner Pass, and the Rhine Valley via Salzburg to Vienna).

Power is supplied from hydro-electric stations in the Bavarian Alps: nearby resources of oil and natural gas are also being exploited in the Alpine foothills and the Inn Valley. Munich's principal industries are textiles, electrical engineering and brewing. The city is rather remote from sources of heavy raw materials and from markets in the Rhineland and North Germany. Hence its factories tend to specialise in producing goods of small bulk but great value, e.g. cameras and optical instruments. Such articles need few raw materials and can be transported great distances with little effect on their selling price.

Munich is also a centre for technical education and its fine buildings and art collections attract large numbers of tourists.

Augsburg was founded by the Romans, and in medieval times became famous as an ecclesiastical and trading centre. It also achieved fame as one of the places where modern banking originated. Today it has important textile and engineering industries.

Regensburg was a key fortress town on the frontier of the Roman Empire. Today it is the main Danube port in South Germany, being at the head of navigation for large vessels and a transhipment point for cargoes passing to and from Austria via the Rhine–Danube Canal.

Ingolstadt, with its very large oil refining and petro-chemical industry (*map p. 99*), has become the main source of fuel and power for industrial growth in South Germany.

Italy via the Alpine passes to the Rhineland, and (ii) from the Danube Valley to the Rineland.

Bavarian Alps
Limestone fold ranges over 4000 m. Very rugged: gorges, precipices, scree-slopes, jagged ridges and pinnacles. Many glacial lakes. Pine forests on lower slopes. Magnificent scenery attracts large numbers of tourists and winter sports enthusiasts. Main resorts: Garmisch-Partenkirchen, Oberammergau, Mittelwald.

THE BAVARIAN ALPS

S. GERMANY:
MAIN TOWNS AND RAIL ROUTES

WEST GERMANY'S 'NEW LOOK'. Until 1948 very little
could be done to make good the damage of the war years. The
rebuilding since then, however, has changed not only the appearance of the country—look again at the photos on pages 84–5—but
also the pattern of its economic life.

Agriculture has been affected in various ways. Before the war
some food eaten in Western Germany was grown on land which
is now behind the 'Iron Curtain'. These supplies have ceased.
Moreover, before the war the population in the area of what is now
West Germany was 42 million. Today it is 61 million, swollen
by 13 million refugees from the Eastern German lands. West
Germany has therefore to feed a much bigger population from a
much smaller area of farmland. To add to the problem, very
many farms in West Germany are far too small (*the average size is
only 9 hectares*) to provide the peasant families who work them
with an adequate income. Hence, since 1949, some 1·2 million
West German farm-workers have left the land to seek more profitable employment in the towns.

West German farmers have therefore been forced to rely on
increased mechanisation; small farmers often hire or share tractors
if they cannot afford to buy them. The Government also enables
them to buy commercial fertilisers at cheap rates, and as a result
the yields of certain crops, notably sugar-beet and grains, have
risen sharply. A start has been made, too, on the merging of
tiny, fragmented farms into larger, more profitable holdings. Today West Germany is about 75% self-sufficient in foodstuffs.

Industry has not only been rebuilt but to some extent remodelled
since 1948, for West Germany has had to make up for the industrial
goods formerly received from the East. These came mostly from
Saxony (*p. 239*) and were mainly engineering products such as
automobiles and machine tools, together with a great range of consumer goods, e.g. textiles and household equipment. The post-
war expansion of West Germany's engineering and consumer
goods industries, notably those of the Ruhr, was made easier by
the flight there of many business-men whose factories had been
seized in East Germany and Czechoslovakia. In some cases, too,
it has been necessary to expand a particular industry to provide
work for refugees. An important example is at the 'New Town'
of Salzgitter, a large proportion of whose 118 000 population are
refugees. Here the former 'Hermann Goering' industrial plant has
been reconstructed and enlarged into a major producer of steel and

engineering goods such as mining and oilfield equipment. The Salzgitter steel mills use iron ore obtained from the sandstones and limestones of Lower Saxony (*map p. 88*), and coking-coal brought from the Ruhr. Lower Saxony now supplies 60% of West German output of home-produced iron ore, and the Salzgitter works (together with an older steel plant at Peine) accounts for about 10% of the country's steel production. Another major development, the growth of great inland oil refineries and petrochemical works, is referred to on pages 99, 100 and 105.

Communications in pre-war Germany tended to run west–east. Major roads, railways and canals stretched across the North German Plain, linking the Ruhr to Berlin, Saxony and the Eastern German lands. The political division of Germany has led to an almost complete breakdown in west–east communications and trade. In West Germany virtually all through movement of goods and passengers is now north–south, i.e. between the North Sea ports and the industrial centres in the Ruhr, the Rhineland and Bavaria. In particular there is now much greater traffic between Hamburg and South Germany. A great new *autobahn* (motor-way) has therefore been constructed to link Hamburg and Stuttgart via Kassel. At the same time canal links between the North Sea ports and the interior are being improved: the Dortmund–Ems Canal has been widened, and navigation improved on the Weser between Bremen and Minden. The principal rail routes affected by the increased north–south traffic are those along the Rhine Valley and the links between Bremen, Hamburg and the Ruhr. These lines are being modernised and electrified. (*For main road, rail and canal routes, see maps pp. 88, 90, 96, 104–5.*)

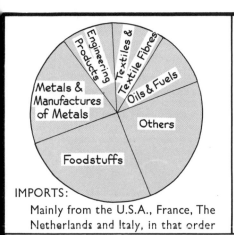

IMPORTS:
Mainly from the U.S.A., France, The Netherlands and Italy, in that order

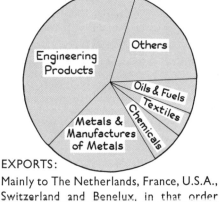

EXPORTS:
Mainly to The Netherlands, France, U.S.A., Switzerland and Benelux, in that order

THE FOREIGN TRADE OF WEST GERMANY

CHAPTER 5

The Low Countries and Luxembourg

THE NETHERLANDS

"An observant foreigner arriving in Amsterdam by air may notice that the control tower at the country's great airport bears a legend to the effect that the airport lies thirteen feet below sea-level. He may also notice a ship gliding by now and then somewhere about the level of the roofs of the airport offices. But it is only a good observer who notices these things. Other visitors from abroad do not make such discoveries, and are soon lost in the daily bustle of one of the most densely populated countries in the world. You would never be able to tell from the appearance of the Netherlands' 13 000 000 inhabitants that more than half of them live and work at six to nine feet below sea-level. They themselves are well aware of the fact, but have no sense of any impending danger. Throughout the centuries Dutchmen have grown up in the sight of water. They are a nation which has become inured to the idea of living in a ceaseless struggle with the forces of nature."*

* A. G. Maris, 'The Ceaseless Struggle', *The Guardian.*

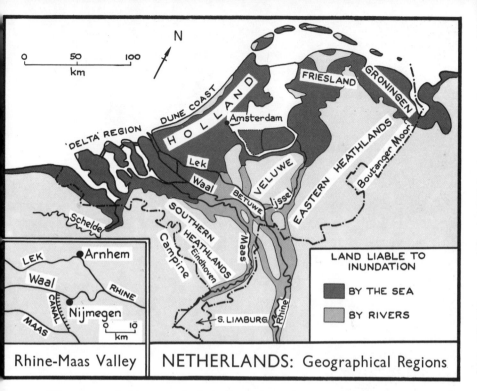

Rhine-Maas Valley | **NETHERLANDS: Geographical Regions**

The scene of this struggle is indicated on the map. (*1*) Estimate what proportion of the country would be flooded by sea or river water during storm tides if there were no protecting dykes. From time to time disasters do occur, as shown by the photograph opposite. On this occasion the tide, whipped up by hurricane-force winds, surged over a large part of South Holland and Zeeland, drowning 1800 people. The land liable to flooding consists mainly of fragments of former sea-bed, together with smaller patches of former lake-floor and alluvial swamp.

Reclamation has gone on since Roman times, new land being formed piece by piece by building dykes round a flooded area and then pumping out the water to make a **polder** (*details overleaf*). Land reclaimed from the sea forms a *marine* polder; land recovered from beneath river water is known as a *riverine* polder. (*2*) Name from the map those regions of the Netherlands which consist almost entirely of polders.

South and east of the polders lie patches of higher, sandy heathland. In the extreme south-east is a fourth region, South Limburg. In spite of its small size it is important because it contains the only worked coalfield in the country.

There are several stages in the formation of marine polderland. The sea-bed to be reclaimed is first sealed off with a **ring-dyke**. Photo **A** shows a length of such a dyke being built in the former Zuider Zee. Its foundations consist of huge quantities of clay, sand and silt dredged from the nearby sea floor. The sides of the dyke are open to erosion by currents and **tidal scour**, and so they are protected by bitumen, plastic seaweed and mattresses (photo **B**), formerly of interwoven willow boughs, now of nylon. These are secured to the clay by stakes, and weighted by thousands of stones.

The top layer of stones is carefully faced with concrete or basalt blocks (photo **C**). If the top of the dyke is sandy, marram grass may be planted and fences built to prevent the sand being blown away (photo **D**). When the ring-dyke is complete, the water inside is pumped out. In the past the pumps were driven by windmills, but today they are mostly powered by steam or diesel engines.

When a marine polder is first pumped dry its soil contains much salt. This is removed within about four years by being dissolved by rain-water and then pumped away. Deep ploughing (photo **E**) improves the soil, and if it is planted with crops the polder will soon look like the one in photo **G**. This is part of the great North-east Polder (map overleaf). Look closely and you will see the outline of the former island of Schokland, now entirely surrounded by fertile farmland. Photo F shows the island before reclamation began.

Polders with soils consisting of sea-clay are extremely fertile and are intensively cultivated with such crops as cereals, sugar-beet, vegetables and flowers. The peaty and alluvial polders are less fertile and are mostly put down in permanent grassland for dairying.

THE COASTAL LANDS consist almost entirely of polders. Between the Hook of Holland and Den Helder, however, the reclaimed land is protected on its seaward side by a natural bulwark—an almost continuous belt of sand-dunes, strengthened in places with stone and concrete. Sheep browse on the dunes, and rain-water is stored in them to supply the towns and also to hold back the salt-water which otherwise seeps inland through the soil and ruins it for farming.

South of the Hook lies the Delta district, until recently a maze of islands and peninsulas separated by long, shallow, tidal inlets. (*3*) Which three major rivers discharge their waters here? This region contains some of the most low-lying yet fertile polders in the Netherlands, and is the most difficult to protect. Notice on the map the great areas which were flooded in the violent storm of 1953. Now a tremendous engineering project—the 'Delta Plan'—is sealing off the region by a series of great dams (*see map*). The project will be completed by 1980 at a cost of £250 million, but this expense will be repaid by the greater security, better harvests (the new dams will prevent salt-water seepage), quicker traffic routes and adequate fresh-water supplies for Rotterdam. It is also planned to develop the region as a boating, fishing and holiday centre.

The 'Delta Plan' poses major engineering problems, but the Dutch can draw on a wealth of experience of such work. So far the most spectacular scheme has involved the reclamation, now

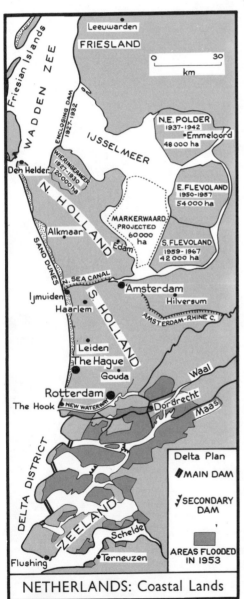

NETHERLANDS: Coastal Lands

A huge petro-chemicals complex is located at Terneuzen. This port ranks fourth in the Netherlands in tonnage of cargo handled.

nearing completion, of the former Zuider Zee. In 1932 a 29 kilometre dam was completed, enclosing the entrance to this shallow gulf. Ring-dykes were then built, and four huge polders have been pumped dry. Their areas and dates of completion, with details of future projects, are indicated on the map.

"The North-east Polder is about 14 ft. below sea-level, and has a dyke 34 miles long running round it. It has one town right in the middle and ten large villages conveniently dotted about. To stand in the town square is to stand in any town square, except that every shop and house looks new. It is all very well to be told that the sea is 14 ft. higher than the dry land of the paving stones but, though the fact is accepted and though the hotel calls itself 'The Hotel Beneath the Sea', the situation seems improbable and just as unlikely as the fact that the earth, so particularly flat in Holland, is round. The polder . . . should really be looked at from its edge, for in the centre nothing can be seen of the water surrounding and threatening it. But on the dyke itself the true position of the polder is shown, for the water laps high on one side of the dyke and on the other the neat fields and straight roads stretch away 4½ yards below the level of the lapping water. . . . Every tree and house and hedge and canal has come into existence since the day when the polder became dry."*

North of Den Helder the sea has breached the dune-belt to form the Frisian Islands. Between the Islands and the coast of Groningen and Friesland lies the shallow Wadden Zee, at low tide a dreary expanse of mud and sand flats crossed by a maze of creeks. At some future date the Wadden may be reclaimed by damming the gaps between the Frisian Islands. Centuries of natural and man-made reclamation from the landward side have already added a strip of territory up to 30 kilometres wide to the Provinces of Groningen and Friesland.

The Coastal Lands form one of the richest farming regions in Europe. (4) Use the information given overleaf to make a *large*, labelled sketch-map with an appropriate title.

Farming in the Netherlands is highly organised and scientific, and crop and milk yields are amongst the highest in the world. Dutch farmers collaborate closely with agricultural research institutions and make full use of such aids as fertilisers, insecticides, special seeds and machinery (*III: 163*).

The high quality of farm produce is further ensured by rigid government inspection, and there are special schools for young people who intend making a career of farming. In recent years, however, the total number of Dutch farm-workers has steadily dwindled, mainly because of the better wages obtainable in industry. Rapid industrialisation since 1945 has changed the face

* 'Farming the Zuider Zee', *The Guardian*.

ZEELAND. Much FERTILE HEAVY CLAY. ARABLE farming very important. Main crops: WHEAT, BARLEY, SUGAR-BEET, POTATOES, grown in rotation.

North and South HOLLAND. INTENSIVE AGRICULTURE, supporting a DENSE RURAL POPULATION. About half of the farmland consists of PEATY and ALLUVIAL CLAY SOILS which are kept in PERMANENT GRASS for DAIRYING. Most milk produced in S. Holland is marketed in liquid form; in N. Holland much CHEESE and BUTTER are made. Famous cheese markets at Edam, Gouda and Alkmaar. Also much MARKET GARDENING in S. Holland, favoured by fine-grained soils and huge demand for fruit and kitchen vegetables from W. European conurbations. Specialised farming districts include:

(i) *Area South of The Hague:*—vast acreage of GLASSHOUSES, producing such crops as TOMATOES, MELONS, CUCUMBERS, EARLY LETTUCE, PEACHES and GRAPES. Over half are exported, chiefly to the Ruhr, Belgium and Britain.

(ii) *Area between Leiden and Haarlem:*—cultivation of BULBS on narrow belt of SANDY SOILS lying just to the east of the dunes. Millions of bulbs (mostly TULIPS) produced annually for export, mainly to Great Britain.

GRONINGEN and FRIESLAND. Rich farming country. HEAVY CLAY SOILS along the coast. INTENSIVE ARABLE farming. Most of region under rotation of crops including CEREALS, GRASS, CLOVER, POTATOES, chicory and mustard. Many Friesian dairy cattle; important DAIRYING industry. Main market centres at Groningen and Leeuwarden.

of the country, and today no less than 41% of the workers are in industry as compared with only 8% in agriculture.

Industry in the Coastal Lands is centred mainly in and near the great ports of Amsterdam and Rotterdam.

Amsterdam first developed in the 13th century as a seaport on the coast of the former Zuider Zee. A dam, built to protect shipping near the point where the little River Amstel entered the sea, gave the city its name. The site of the original 'Dam' is now a broad open place in the heart of the city. Notice, too, the remarkable series of semi-circular canals: these were built as the city gradually spread southwards across reclaimed swamp at the edge of the Zuider Zee. Through the centuries the port developed into a flourishing commercial, financial and industrial centre. In particular it became the main focus of Dutch colonial trade. (5) Which of the present industries probably originated from the importation of colonial raw materials?

Since 1876 Amsterdam's main outlet to the North Sea has been via the North Sea Canal. This great waterway has enabled the port to expand its entrepôt and industrial activities in spite of the sealing off of the Zuider Zee. There is also an

Main Industries of
AMSTERDAM

Metal working
Engineering
Shipbuilding
Flour-milling
Diamond cutting
Sugar-refining
Tobacco processing
Vehicle assembling
Oil-seed crushing
Manufacture of:—
  Cocoa and chocolate
  Clothing

AMSTERDAM AND THE NORTH SEA CANAL

important link to the Rhine via a new Amsterdam–Rhine Canal, opened in 1952. Amsterdam is now a magnificent city of 1 044 000 people, and its industrial suburbs reach along the North Sea Canal as far as *Ijmuiden*. The latter port has a large iron and steel works—the only one in the Netherlands—using sea-borne coke and iron ore. It is also the largest Dutch fishing port.

Rotterdam is the largest port in Europe. It has existed for more than 600 years, but became a major port during the 19th century. Its rapid growth and prosperity in modern times have gone hand in hand with the increase in population, industry and trade of its hinterland. The latter, due to Rotterdam's geographical position, is enormous. A complex pattern of navigable waterways, both natural and man-made, converges on the city from industrial areas in the Netherlands, Germany, Austria, Switzerland, Belgium and Eastern France (*see map*). Thus, in addition to becoming the largest commercial port of the Netherlands, Rotterdam has become an entrepôt for a large part of Western Europe, to which it is linked chiefly by the River Rhine and its tributaries. Above all, Rotterdam is the main port for the Ruhr industrial area (*p. 97*).

At the heart of Rotterdam lies a maze of docks, warehouses and harbour installations. About 75% of the 150 000 000 tonnes of cargo handled there annually represents transit trade for countries outside the Netherlands. The chief imports are of bulk raw materials such as petroleum, iron ore and wheat, whilst outgoing cargoes consist mainly of manufactured goods. ((6) *Suggest reasons for this trade pattern.*) Rotterdam is a major industrial centre with port industries including petroleum refining, flour-milling and the manufacture of chemicals. It is also one of the world's main shipbuilding and marine engineering centres, the second Dutch fishing port and a naval base. Although the sheltered estuary of the New Maas gives Rotterdam a natural harbour, the port fights a constant battle against silting.

Since 1872 Rotterdam has been linked to the North Sea by

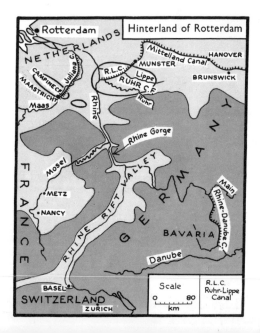

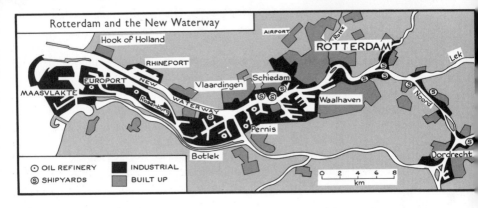

Rotterdam and the New Waterway

the New Waterway, a fine deep ship-canal able to take very large
ocean-going vessels. Great developments are at present taking
place on the banks of the Waterway, part of the phenomenal
growth of Greater Rotterdam. In particular an international port
with first-class facilities has been constructed at Europort, at the
seaward end of the island of Rozenburg (*see map*). By 1973
Europort will cover a total area of 5000 hectares, two-fifths of them
reclaimed from the shallow bed of the North Sea. The decision
to build Europort was made in the light of three major trends in
European trade:—(i) an enormous increase in demand for refined
petroleum products, coupled with the use of very large super-
tankers; (ii) the building of bulk-carriers of up to 250 000 tonnes
to carry cargoes such as iron-ore, grain, oil-seeds and solid fuels
and (iii) the rapid spread of container traffic for general merchan-
dise. The deep-water berths of Europort are specially designed
to take these mammoth vessels and the quays are equipped with
devices to ensure swift handling and despatch of cargoes. The
unique location of the new docks at 'the gateway to Europe' gives
them easy access by road, rail, pipeline and canal to a rich industrial hinterland containing 160 million people, all within a radius
of 500 kilometres.

The new container port has proved so successful—container
services link Rotterdam to places as far afield as northern Italy—
that a dock extension equipped to handle container ships, luxury
liners, ferries, hovercraft and hydrofoils has been built on the
north bank of the Waterway at Rhineport (*see map*). Europort
also has an enormous oil refining and petro-chemicals industry,
with a combined annual capacity of over 100 million tonnes: on
average a loaded supertanker of over 150 000 tonnes arrives at
Europort to discharge its cargo every second day. (7) Draw a
sketch-map to show that crude oil is also sent by pipelines to

Europort oil terminal, with the New Waterway beyond

refineries at Cologne, Frankfurt, Amsterdam, South Limburg (*see overleaf*) and Terneuzen (*map, p. 112*). Other major projects in Europort include a blast furnace and steel mill (using imported ores and coke), and a chemical fertilizer works. The volume of goods now passing annually through Europort exceeds that of the total trade of the United Kingdom.

Other important towns in the Coastal Lands are:—

The Hague (Den Haag—popn. 750 000), primarily a political and administrative centre, being the home of the Dutch Parliament and of the Permanent Court of International Justice. The city has spread towards the coast and now includes the fashionable seaside resort and fishing port of Scheveningen. Many people who live at The Hague travel daily to work in the Rotterdam conurbation. Although it is mainly a residential city, the number of banks, offices and industries in The Hague has increased greatly in recent years. Industries include printing, paper-making, clothing and food-processing.

Delft (72 000) and Leiden (158 000), both flourishing medieval trading centres. To-day the quaint charm of their older buildings attracts large numbers of tourists. There are also light industries such as cigarette-making and distilling.

Dordrecht (81,000), once the leading port of the Netherlands, but silting and the increasing size of ships caused it to be eclipsed by Rotterdam in the 19th century. Now its link to the sea via the Maas has been deepened and Dordrecht has industries such as shipbuilding, marine engineering, chemicals and oil-seed crushing.

THE INTERIOR NETHERLANDS

(a) *The Rhine–Maas Valley* is a broad lowland of riverine polders forming a distinct region across the centre of the country. The maps on pages 109 and 116 show the principal **distributaries** of the River Rhine: the Waal, Noord and New Waterway carry the main water discharge. In addition to the main channels shown on the map, many smaller distributaries and artificial cuts, the most important of which is the Maas–Rhine Canal, have been made for drainage and to facilitate navigation. For centuries this region has been afflicted by devastating floods, for the gradient of the rivers is very gentle and deposition of alluvium has raised their beds and banks above the level of the adjoining countryside. An additional hazard results from the shrinkage and consequent lowering of the reclaimed polders as they become dry. Today the polders are protected by massive dykes: in particular a huge wall has been built to keep the Maas separate from the Rhine.

In spite of the enormous volume of barge traffic plying to and from Rotterdam on the principal rivers, the Rhine–Maas Valley has hitherto been a rather remote and sparsely peopled region. The constant flood danger has deterred settlement and the maze of waterways has impeded land communications. Now the Government is attracting would-be farmers to this 'pioneer' region by offering financial aid.

Most of the riverine polders have moist, clay soils, and dairying is the main farming activity. In places where the soil is lightened by patches of sand, as in the Betuwe (*map p. 109*), much orchard fruit is grown.

The scattered villages and farms cling for protection to the main dykes. The only important towns in the region are *Arnhem* and *Nijmegen*. These stand on slightly higher ground above flood level and for centuries have been crossing places of the Rivers Lek and Waal respectively. Their vital strategic position was demonstrated in 1944, when Allied airborne forces made a massive but ill-fated attempt to seize the Rhine crossings from the retreating Germans. Both towns have developed as important river ports, and both have varied engineering and processing industries which rely on waterborne raw materials. These include the manufacture of rayon and chemicals at Arnhem, and sugar-refining, brick-making and distilling at Nijmegen.

(b) *The Heathlands* are located on extensive, undulating tracts of sands and gravels (*map p. 109*). In the north and east (*see*

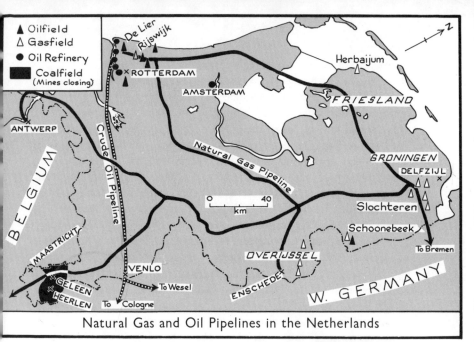

Natural Gas and Oil Pipelines in the Netherlands

p. 88) these deposits are of glacial origin, but the Campine sands were laid down by ancient forerunners of the Rivers Maas and Rhine. In some places there are considerable areas of bare sand, with dunes reminiscent of a desert landscape. Elsewhere are great stretches of heather, birch-scrub and rough pasture. Much of the heathland is too dry for cultivation, for the rain-water rapidly soaks away through the unconsolidated sand. In some parts, however, where drainage is impeded by saline 'pans', peat-bogs and meres have developed, e.g. the Bourtanger Moor.

Since the late 19th century considerable progress has been made in reclaiming the Dutch heaths. Peat-moors around Groningen, for example, have been drained and now yield good crops of rye, oats, sugar-beet and potatoes. Many coniferous forests have also been planted (especially in the Veluwe) and poor pasture land improved by the sowing of special grasses.

For centuries a neglected and sparsely-peopled region, the Dutch heathlands have recently attracted great attention following the discovery there of petroleum and immense quantities of natural gas. Since 1944 petroleum has been pumped from below ground at Schoonebeek on the German frontier (*map above*). Together with the output of other wells around Rijswick (between Rotterdam and The Hague), about 2 m. tonnes of petroleum, i.e. one-quarter of the total Dutch requirements, are now produced annually. One of the world's largest known fields of natural gas lies in the provinces of Groningen and Overijssel. Deposits amounting to at least 1 800 000 m. cubic metres have been discovered there since 1960. This phenomenal find is equivalent to 11 000 million tonnes of coal, and successful test borings are still being made in Friesland, Drenthe and N. Holland. The gas is being used to develop industry in the Northern Netherlands, for long a problem area of high unemployment; aluminium smelting works and fertilizer factories have been established, for example, at Delfzijl. Hitherto the only important industrial activities in the heathlands have been (i) the processing of agricultural produce, e.g. flour-milling, brewing and the manufacture of starch, alcohol, glucose and strawboard) (8) *from which raw materials in each case?*) at Groningen; (ii) the manufacture of cotton textiles—an old-established industry—in such towns as Hengelo and Enschede (*map p. 88*) and (iii) light engineering, notably in the vast Philips electrical factory at Eindhoven. (*Map p. 109.*)

Groningen gas is also sent by pipeline to all parts of the Netherlands (*map, p. 119*), where total gas consumption in homes and factories has risen to some 17 000 million cubic metres p.a. This represents about one-quarter of the total energy requirements of the country. Gas is also exported in huge and increasing quantities to West Germany, Belgium and France.

(c) South Limburg is rather remote from the rest of the Netherlands, but it contains some of the country's most fertile loam soils, probably developed from deposits of loess (*p. 12*). Together with some sands and marls they form a 'blanket' over the chalk plateau of which South Limburg is chiefly composed. Much sugar-beet, potatoes and fruit are grown, mostly on mixed farms which also support dairy cattle and pigs.

The coalfield (*map, p. 119*) although concealed and difficult to work, has been of great value since large-scale modern mining commenced there in the 1900's. Now that the huge Groningen gasfield provides a much larger and cheaper alternative source of power the coalfield is being run down. Only five of the original twelve mines remain open and all of these are to be closed in the next few years. The considerable chemical and engineering industries of Geleen, Heerlen, Sittard and Maastricht, formerly dependent upon local coal, have switched to piped gas and oil. In fact the use of Groningen gas, both as a fuel and a raw material, is allowing a big expansion of South Limburg's petro-chemical industries.

With an average of 378 persons per km² the Netherlands is the most densely populated country in the world. Population pressures are especially high in and around Utrecht, Amsterdam and Rotterdam (3023 per km²), so that the Dutch are faced with acute problems of urban growth—suburban sprawl, traffic congestion, loss of farmland, pollution, water supply, disfigurement of the countryside, and so on. The map shows that the cities of Rotterdam, The Hague, Leiden, Haarlem, Amsterdam and Utrecht have virtually coalesced to form a ring-shaped conurbation which the Dutch call *Randstad Holland*. The damp, peaty, but pleasant rural region around which the Randstad towns are grouped form a ' green heart ' which planners aim to preserve at all costs. This ' heart ', as well as being an oasis of green in a growing urban ' desert ' is vital to the Dutch economy because of its bulbfields, market gardens and glasshouses.

The problems are immense, for the Dutch rate of natural increase (1·16% p.a.) is the highest of any country in western Europe and the present population of 13 million is expected to reach 18 million by the end of the century. Town populations are also being increased by a massive movement of labour out of agriculture into industry, the agricultural work force having fallen by about 300 000 since 1947. The immediate need is to build houses for over a million people in the western Netherlands (Holland and Utrecht), but builders have to compete for land with (i) market-gardeners anxious to increase the area under glass; (ii) industrialists wishing to expand their premises and build new factories and (iii) port operators designing new warehouses and dock installations.

To avoid making inroads into the ' green heart ', future growth will be permitted only in certain well-defined directions. The main areas scheduled for development are (a) the main transport routes leading away from the Randstad, e.g. from IJmuiden northwards towards Alkmaar; (b) a narrow strip of land running north-east from Amsterdam through the new South Flevoland Polder and East Flevoland towards Groningen and (c) the island of Voorne in the Delta District, where a new town Grevilingenstad is projected to absorb 250 000 ' overspill ' population from Rotterdam. In addition several cities remote from the Randstad are to be developed and expanded, in some cases to over 200 000. Finally, to preserve a semblance of green countryside within the Randstad, a rural buffer-zone 4 kilometres wide is to be retained between all the existing large cities.

One disadvantage of this trend is that South Limburg forms part of that shrinking portion of the Netherlands which is still predominantly rural. The pleasant open countryside of the southern and eastern provinces forms a welcome contrast to the heavily urbanized landscape of North and South Holland. Details of Dutch land-use problems are given in the above table.

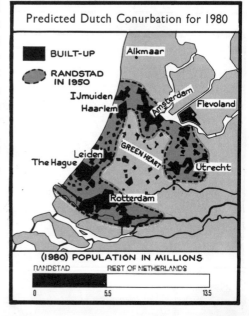

Predicted Dutch Conurbation for 1980

BUILT-UP

RANDSTAD IN 1950

Alkmaar
IJmuiden
Haarlem
Amsterdam
Flevoland
Leiden
The Hague
GREEN HEART
Utrecht
Rotterdam

(1980) POPULATION IN MILLIONS
RANDSTAD · REST OF NETHERLANDS
0   5.5   13.5

Hothouses near Brussels.

The Belgian Coast near Blankenberge.

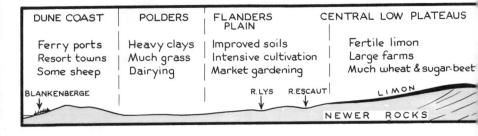

DUNE COAST	POLDERS	FLANDERS PLAIN	CENTRAL LOW PLATEAUS
Ferry ports	Heavy clays	Improved soils	Fertile limon
Resort towns	Much grass	Intensive cultivation	Large farms
Some sheep	Dairying	Market gardening	Much wheat & sugar-beet

BLANKENBERGE · · R.LYS R.ESCAUT LIMON

NEWER ROCKS

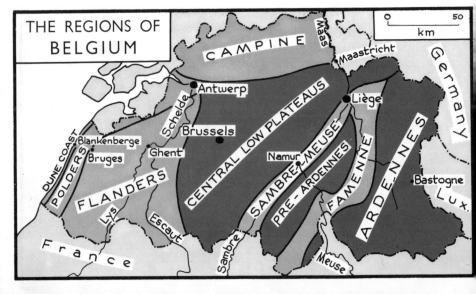

THE REGIONS OF BELGIUM

CAMPINE

Maas

Maastricht

Germany

Antwerp

Schelde

Brussels

Liège

Blankenberge

Ghent

CENTRAL LOW PLATEAUS

Bruges

DUNE COAST

POLDERS

FLANDERS

Namur

SAMBRE-MEUSE

PRE-ARDENNES

FAMENNE

ARDENNES

Bastogne

Lys

Escaut

Sambre

Meuse

France

Lux.

0 — 50 km

The Meuse Gorge at Dinant.

AMBRE-MEUSE VALLEY	CONDROZ 'PRE-ARDENNES'	FAMENNE DEPRESSION	ARDENNES Damp, open moorland
coalfield	Forested ridges	Mostly shales	Desolate
heavy industry	Sandstone, shale	Mixed farming	BASTOGNE
iron & steel	& limestone		
chemicals			ANCIENT CRYSTALLINE ROCKS

CARBONIFEROUS ROCKS

BELGIUM

The main facts about the geography of Belgium are shown in the map, diagrams and photographs on these two pages. Study them carefully and then (*10*) write a geographical account of a journey from Blankenberge to Bastogne. Pay particular attention to the geology, scenery, agriculture and other activities in the regions through which you would pass. (*11*) Which is the only region in Belgium you would *not* cross on this journey?

Although Belgium is a small kingdom, with a total population of only 9·6 million, it contains many contrasts of great geographical interest. The flat polders and fertile, undulating plains of northern Belgium stand in striking contrast to the wooded hill-slopes and barren moorlands in the south. Intensively cultivated

and densely populated farmlands lie side by side with, and sometimes encircle, coalmining districts and industrial towns.

Another contrast is afforded by the Belgians themselves. The
inhabitants of northern Belgium are Flemings—a Nordic people
who speak a language similar to Dutch, and who are predominantly
Roman Catholic and Royalist; the Southern Belgians are Walloons,
a people of Celtic origin who speak a variant of French, and who
are mainly Protestant and Republican. Relations between Flemings and Walloons are poisoned by a jealous animosity which at
times leads to outbreaks of physical violence. To symbolise equal
standing for the two languages and peoples, such items as coins,
bank-notes, official documents and even some place-names are
designated in both French and Flemish e.g. Anvers and Antwerp,
Liège and Luik. The photo shows a street name-plate in Brussels.

As well as internal discord, the Belgians have had to contend
with outside aggression. Belgium's position at the narrowest part
of the North European Plain gives it great strategic importance, so
that in both World Wars it was invaded and devastated by oppos-
ing armies.

FARMING IN BELGIUM is intensive. In their natural state
the soils in many parts of the country were rather sterile, but centuries of meticulous cultivation and heavy manuring have made
them highly productive. As a result the Belgians are able to produce about 80% of their food requirements, and maintain a high
standard of diet in spite of the fact that they have one of the
greatest densities of population in the world. The table shows
that many Belgian farms (notably those in Flanders) are remark-

ably small: (*12*) what is the average farm size? In certain circumstances such small plots would hamper agricultural production, but the tiny Belgian holdings are primarily market

BELGIUM: FARM FACTS	
Proportion of working population in farming	6%
Total area in farmland (hectares)	1 825 600
Total number of farms	990 913
Average size of farms	? ?
No. of farms less than 1 hectare	722 582
No. of farms 1–5 hectares	47 599
No. of farms 5–50 hectares	102 250

gardens, owned by peasants who work over the ground inch by inch using hand-spade methods. As a result arable crop yields are amongst the highest in Europe and it is reckoned that a family of eight is able to subsist comfortably on a holding of 5–6 hectares. Much of the work is done by women and children, for the men often have additional jobs in local factories.

Soils in Belgium vary considerably from place to place and there are corresponding contrasts in types of farming. Details of the latter are given below. (*13*) Make a large copy of the map on page 122, using different colours for each region: then add

THE FARMING REGIONS OF BELGIUM (See Map p. 122)

The COASTAL DUNES are of little value except as SHEEP PASTURES.

The POLDERS form a belt from 13–16 kilometres across, immediately behind the dunes. They consist of HEAVY DAMP CLAY SOILS, kept mainly in GRASS, and supporting an important DAIRYING industry. Fodder crops such as oats and sown grass are also grown. Because of the danger from flooding, most farms in the polders are located where slightly higher patches of sand rise above the former marshes.

INTERIOR FLANDERS is very densely populated, with more than 400 people per square kilometre. This sandy region is one of the most INTENSIVELY FARMED districts in the world. Great quantities of MARKET GARDEN PRODUCE, together with rotation crops such as POTATOES and SUGAR-BEET are produced, mostly on VERY SMALL FARM PLOTS. Each holding also usually has two or three dairy cattle. Most peasant farmers in Flanders send produce to market in nearby towns such as Bruges and Antwerp. FLAX has been grown for centuries on the damp heavy clays of the Lys and Escaut valleys.

THE CENTRAL LOW PLATEAUS lie between the Rs. Lys and Meuse. This region has an extensive covering of LIMON (p. 12) and its brown loam-soils,

heavily fertilised, yield fine crops of WHEAT and SUGAR-BEET. The population density is less than that of Flanders, for there are many LARGE PROSPEROUS FARMS and fewer intensively cultivated small-holdings. Around Brussels MARKET GARDENING is very important (*photo p. 122*).

SOUTH OF THE MEUSE there is a contrast between the CULTIVATED VALLEYS in the Pre-Ardennes and the Famenne Depression (*map p. 122*) and the DAMP OPEN MOORLAND of the high Ardennes. The cultivated lower land yields crops of oats, clover, potatoes and rye, and dairy cattle are reared. On the poorer upland pastures beef cattle are reared, but are sent to the polders for fattening.

THE CAMPINE is an undulating sandy plain—a region of heather, swamps and pine forests very similar to the heathlands of the Netherlands (p. 118). It contains the poorest farmland in Belgium, but during the past century successful attempts have been made to improve some of its soils for farming. Now DAIRY CATTLE are reared on IMPROVED PASTURES sown with drought-resistant grasses, and SOME ARABLE crops (e.g. potatoes and sugarbeet) are grown where soils have been heavily fertilised and marled.

labels from the farming notes to make a 'Farming Map of Belgium'.

INDUSTRY IN BELGIUM has been important for centuries. In the Middle Ages Flanders was already densely populated and a main centre of European civilisation. Towns like Bruges, Ghent and Brussels became famous for their great annual trade fairs; the main goods exhibited were textiles, especially woollens and lace, made from locally produced wool and flax. Other textile centres grew up in the valleys of the Ardennes, where wool and water power were readily available. Some metal-working towns such as Liège, owed their early development to supplies of mineral ores from these Hercynian uplands

Modern Belgium is still an important industrial country. Not only have many old-established towns maintained, expanded and diversified their industries, but whole new industrial regions have emerged, notably (i) along the Sambre–Meuse 'coal-furrow', (ii) in and around Antwerp and (iii) in Eastern Belgium along the banks of the Albert Canal and in the Campine.

THE SAMBRE-MEUSE VALLEY is about 160 km long and from 5–16 km wide. In this narrow trough are crowded $2\frac{1}{4}$ million people—one-quarter of the entire Belgian population. They live and work in a series of coalfield industrial towns (*see map overleaf*) and sprawling 'satellite' villages. Some of the towns, notably Liège, have a long history of metal-working, but large-scale industrial expansion came with the development of the coalfield in the 19th century. Now the Sambre–Meuse region is a major centre of heavy industry, with emphasis on iron and steel manufacture, non-ferrous metal-working, engineering, chemicals and glass-making.

The coalfield is diffi-cult and dangerous to work, for the coal-seams have been badly con-torted and fractured by

Sambre–Meuse Coalfield Number of Collieries	
1875	175
1939	77
1955	63
1965	51
1971	5

Near MONS there is a single mine shaft 372 m deep, which passes through the same coal-seam six times.

Contorted and Fractured Coal-seams

movements in the Earth's crust. The diagram shows the sort of conditions met with below ground near Mons. Explosions of 'fire-damp' are commonplace and costs are increased by the great depth of many of the mines (over 1100 m near Mons). In addition, some of the coal is so shattered that it has to be compressed into briquettes at the pit-head before it can be marketed. All easily accessible coal had been removed by the end of the 19th century, and many of the early mines now lie derelict: indeed the whole of the field round Namur is exhausted. (*Note the decline in the total number of collieries during the past century*.) In spite of these limitations the various bituminous, coking and semi-anthracitic coals produced in the Sambre–Meuse field explain the industrialisation of Southern Belgium.

"It is obvious that much of the southern coalfield is an 'old' industrial area from its appearance. Derelict collieries, overgrown spoil-banks, a chaos of pit-shafts, blast-furnaces and steel-works, chemical factories, long rows of small, drab, gardenless dwellings built in irregular rows—all these are typical of the crowded and haphazard industrial development of the nineteenth century. However, not all the southern coalfield is like this; new housing estates, and (particularly in the western part of the field) more open industrial villages, small-holdings and farmland intermingle in a manner characteristic of so many parts of Belgium. To the north of Mons, away from the cramping bounds of the Meuse valley, extend the fertile limon-covered arable lands of the Hainaut plateau. But elsewhere the very concentration of industrial activity along the narrow line of the Sambre-Meuse valley leaves little space for planned development."*

Further details of industries in the Sambre–Meuse Valley are given overleaf.

* F. J. Monkhouse, *A Regional Geography of Western Europe*, Longmans.

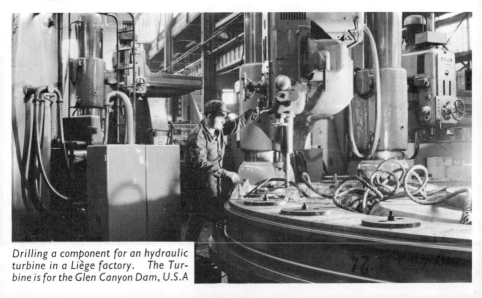

Drilling a component for an hydraulic turbine in a Liège factory. The Turbine is for the Glen Canyon Dam, U.S.A

BRUSSELS (Over 1M), capital and main industrial centre, originated in 6th century as defensive site on small island in R. Senne. Old 'lower town' is industrial and commercial. New 'uppertown' spreading eastwards on adjacent plateaus is mainly residential. Industries include metallurgy, chemicals, textiles, paper and furniture. Routes converge on Brussels from Ostend/ Antwerp and Netherlands/ Ruhr/ 'Coal Furrow' and S. Belgium/Paris and Lille. A major inland port, reached by ships up to 3000 tonnes via Willebroek Canal, and by barge from Charleroi.

ANTWERP* (About ½m.). On deep, sheltered, navigable estuary of W. Scheldt. Focus of inland waterways from (i) Flanders, via Scheldt and tributaries; (ii) Sambre–Meuse area via Albert Canal; (iii) Rhineland, via new canal to Dordrecht. World's third largest port, handling four-fifths of Belgium's trade and much entrepôt traffic. Main imports: petroleum, ores, foodstuffs, 'colonial goods', timber, coal. Main exports: manufactured goods, esp. steel, chemicals, glass, textiles. ((14) Suggest likely sources of these goods.) Main industries: vegetable oil, petroleum and sugar refining; manufacture of soap, margarine, chocolate, rubber goods and chemicals ((15) from what imported raw material(s) in each case?); metallurgy and engineering; vehicle assembly.

(16) Draw labelled sketch maps showing Brussels and Antwerp as route centres.

GHENT (200 000). Ancient industrial and commercial city at junction of Lys and Upper Scheldt. Ship-canal link to sea at Terneuzen. Dominant centre of all Belgian textile industries except woollens.

Large petro-chemical, fertilizer, metallurgical, glass, food-processing and paper factories, both in city and along ship-canal to north. Magnificent buildings of old inner town attract many tourists.

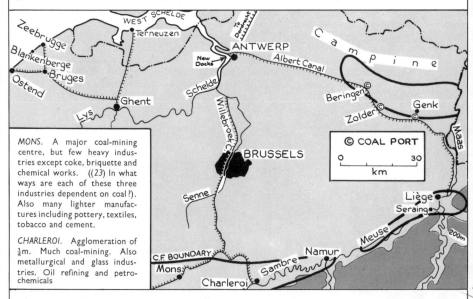

MONS. A major coal-mining centre, but few heavy industries except coke, briquette and chemical works. ((23) In what ways are each of these three industries dependent on coal?). Also many lighter manufactures including pottery, textiles, tobacco and cement.

CHARLEROI. Agglomeration of ¼m. Much coal-mining. Also metallurgical and glass industries. Oil refining and petrochemicals

NAMUR (50 000). Ancient fortress and market town ('The Gateway to the Ardennes') at Sambre/Meuse confluence. Local coal now exhausted, replaced by waterborne supplies from elsewhere on coalfield. Main industry: heavy engineering. Also cement, glass, paper, soap.

LIÈGE (250 000). Chief industrial centre in 'Coal Furrow'. Metal-working, based originally on charcoal-smelted Ardennes ores, developed during 19th century into large-scale steel and heavy engineering industries which now produce (e.g.) boilers, bridges, diesel engines and locomotives. Other industries include light engineering (e.g. hardware, motor-cycles, small arms), zinc smelting, chemicals, glass, tyres. (17) Suggest from the map and atlas why Liège is a major focus of communications.

THE CAMPINE is an undulating sandy plateau, for centuries a 'negative' region of heaths, sand-dunes and shallow marshy depressions. Recent attempts to reclaim the region for farming and forestry are referred to on page 125. Of far greater significance for the prosperity of Belgium is the industrial development which has occurred there during the past half-century. The concealed Campine coalfield was first worked in 1901, since when coalmining has greatly expanded. Although the seams are deep below ground (between 500 m and 1000 m) and costly to work they contain coking-coals of great value to Belgium, for supplies from the southern coalfield are running out.

(18) What proportion of the Belgian coal output now comes from the Campine? The main mining centre is Genk: (19) what was the percentage increase in the population of this town between 1910 and 1971? About one-third of the present inhabitants of Genk are foreigners, mostly Italians, for labour is scarce in this

Population of Genk	
1846	1 776
1910	3 422
1956	43 618
1971	57 375

part of Belgium and foreign immigrants have been recruited to work in the mines.

Coalmining in the Campine has stimulated the growth of industry in Eastern Belgium. Other reasons include:—

(i) the deliberate siting *of dangerous and noxious industries (e.g. chemicals and explosives) in remote, unpopulated heathland;*

(ii) cheap, level land *with plenty of room for factory growth;*

(iii) cheap transport *afforded by the Albert Canal* (see map);

(iv) government encouragement *to industrialists, in the 1920's, to move into the Campine to try to make good the immense damage suffered in the 'Coal Furrow' during the trench warfare of 1914–18.*

Today the Campine is one of Belgium's leading industrial regions, with many large, modern plants concerned mainly with the manufacture of chemicals and glass and with the refining of zinc and other non-ferrous metals. Most of these concerns are sited on the banks of the Albert Canal. This is now one of the busiest inland waterways in Europe; it carries a great volume of through traffic between Antwerp* and Liège as well as the raw materials and finished products of factories along its banks. Some 4 m. tonnes of coal are also shifted along the Canal from specially constructed ports on the coalfield. (20) Name them.

* Since 1960 the volume of cargo handled at Antwerp has trebled, the area covered by industrial sites has increased five-fold and the port's quay length has doubled. New docks have been excavated (see map) and the oil-refining, petro-chemical and vehicle assembly industries have greatly expanded. This spectacular growth is mainly the result of Antwerp's central location in the Common Market, to which it is linked by excellent road, rail and canal networks. Furthermore, room can easily be found for new factories by draining marshland adjacent to the port.

THE FLANDERS COAST has been tremendously changed, during the past half-century, by the rapid growth of the tourist industry. The largest and most popular resorts are Ostend and Blankenberge, but the entire dune-belt between the French and Dutch frontiers is a chaos of villas, caravan-parks, hotels and holiday camps. During the summer months tens of thousands of holiday-makers, many of them British, visit Northern Belgium and the money they spend brings prosperity to the local people.

There is also a flourishing fishing industry, based mainly on the ferry-port of Ostend. This port has fish-curing, canning and refrigerating works, fish fertiliser factories and shipbuilding yards which specialise in making trawlers and other small craft. Elsewhere along the coast there are chemical works at Nieuwport and Zeebrugge, and the latter port has a coke-oven plant using sea-borne raw materials. Zeebrugge lies at the seaward end of a ship-canal from Bruges, a market-town and tourist centre with small-scale industries including textiles, food-processing and engineering.

(*21*) Draw a labelled sketch-map of the Flanders coast, including all the places and industries mentioned in this paragraph. Add a labelled arrow to show that Ostend has important road and rail links with Brussels. Give your map an appropriate title.

LUXEMBOURG

The Grand Duchy of Luxembourg grew up as a small independent state around the natural fortress of Luxembourg City. The northern third of the Duchy forms part of the Ardennes uplands. Here, as in the Belgian Ardennes, the rocks are old and

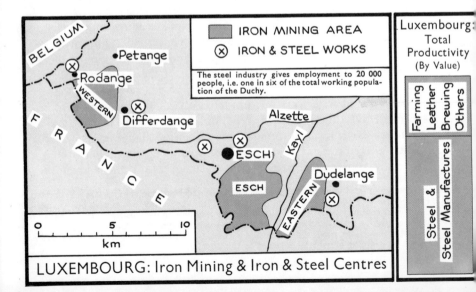

LUXEMBOURG: Iron Mining & Iron & Steel Centres

hard and the soils thin and infertile, and pastoral farming pre-dominates. Farther south the land is lower, the climate milder and soils are more fertile. This is the Bon Pays ('good land'), a district of sandstone and limestone escarpments with intervening clay vales. The landscape of Southern Luxembourg is dotted with prosperous mixed farms, where dairy cattle are reared and cereals, vegetables and orchard fruit are cultivated.

The chief wealth of Luxembourg, however, is obtained from below ground, for the northern tip of the great Lorraine iron-ore field (*p. 67*) just crosses into the Duchy. This fortunate chance has enabled Luxembourg, whose total population is only 338 500, to become the world's thirteenth largest steel producer. The main iron-working districts and iron- and steel-making centres are indicated on the map. The main steel *using* (i.e. engineering) works are in Luxembourg City, where a wide range of products is made including bridges, cranes, factory machinery, rolling stock and agricultural implements. The Duchy's land-locked position adds to transport costs and so emphasis is placed on making high-grade export goods such as ferro-alloys and very fine wire for radial tyres. Transport problems have eased considerably since the canalization of the Moselle (*see page 92*).

Despite recent attempts to establish more varied industries (*see right*) nearly 70% of Luxembourg's total exports consist of steel and steel manufactures.

Some Newer Industries in Luxembourg
Aluminium smelting
Chemicals   Clothing
Tyres   Wines
Household equipment

In 1948 Belgium, The Netherlands and Luxembourg formed a Customs Union called *Benelux*. Customs duties on trade between these countries were abolished and a common tariff was imposed on goods imported from countries outside the Union. The main object was to increase the volume of internal trade by assuring industrialists and farmers of a large home market for their pro-ducts. The Union has been a marked success and has helped to solve several difficult economic problems, e.g. (i) the undue reliance of Luxembourg on its iron and steel industry; (ii) unem-ployment in some of Belgium's older, declining industries, such as some branches of textiles; (iii) shortages in The Netherlands resulting from that country's rapidly increasing population. The alliance of Dutch agriculture and light industry with the heavy industries of Belgium and Luxembourg has produced a more balanced overall economy.

CHAPTER 6

The Alpine Lands: Switzerland and Austria

THE map on page 8, or an atlas, shows that the fold ranges of the Alps form an immense arc from Northern Italy to the Danube Valley. (*1*) Estimate the length and average breadth of these ranges and name each country through which they run. Strictly speaking the term 'Alpine Lands' includes parts of all these countries, but usually it refers just to Switzerland and Austria. (*2*) What approximate proportion of these two countries is occupied by the Alps?

Their structure and relief are strikingly similar, and scenes of breath-taking grandeur such as that shown opposite are common to both. In the high Alpine valleys Swiss and Austrians also follow much the same way of life, with pastoral farming, forestry and tourism as the main sources of income. From the map and from the table below, (*3*) state which country has (*a*) the smaller proportion of arable land; (*b*) the higher density of population; (*c*) the smaller reserves of mineral wealth. Yet (*4*) which nation, on average, is the more prosperous?

The reasons for the different levels of income are largely bound up with the very different political histories of the two countries. Switzerland's successful policy of independence and neutrality has enabled agriculture, industry and trade to develop almost untouched by the ravages of war. Austria, on the other hand, is a remnant of the former Austro-Hungarian Empire which was dismembered after the First World War. Between 1918 and 1938 recovery was delayed by political unrest. In 1938 Austria became part of Germany and by 1945 was sharing that country's devastation (*p. 84*). In 1955 Austria re-emerged as an independent, neutral state, and though the troubled history of recent decades has left its mark, great efforts are being made to develop the country's industries and to make farming more efficient.

	Switzerland	Austria
Population	6 220 000	7 370 000
Area	41 470 km²	84 178 km²
Area of arable land	2 655 km²	16 167 km²
Number of inhabitants per acre of arable land	? ?	? ?
National Income	£7 840 m.	£5 181 m.
National Income per head of population	? ?	? ?

The Lauterbrunnen Valley, Bernese Oberland.

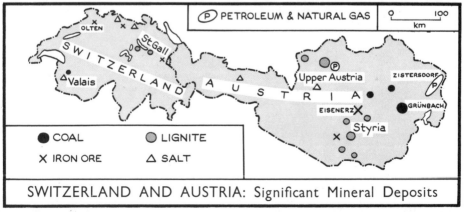

Ⓟ PETROLEUM & NATURAL GAS

0 — 100
km

OLTEN

St Gall

SWITZERLAND

Valais

AUSTRIA

Upper Austria

Ⓟ

ZISTERSDORF

Ⓟ

EISENERZ

GRÜNBACH

Styria

● COAL   ● LIGNITE

✕ IRON ORE   △ SALT

SWITZERLAND AND AUSTRIA: Significant Mineral Deposits

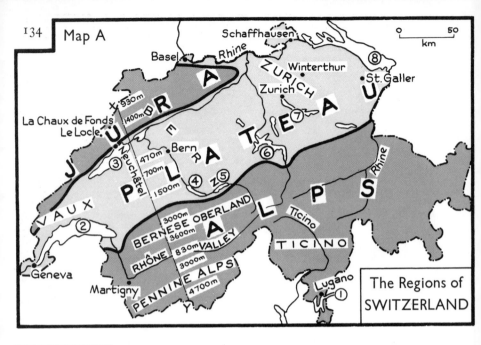

Map A

The Regions of
SWITZERLAND

SWITZERLAND

Population of largest towns	
Zurich	657 000
Basel	354 000
Geneva	293 000
Bern	251 000
Lausanne	207 000

These maps show many of the main facts about the geography of Switzerland. Study them carefully and then (5) with the help of an atlas do the exercises and answer the questions opposite.

BERN—typical Swiss buildings in an old part of the city.

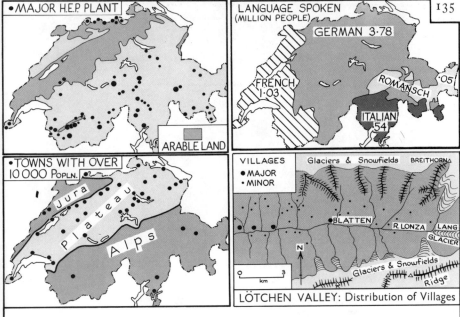

MAJOR H.E.P. PLANT

ARABLE LAND

LANGUAGE SPOKEN (MILLION PEOPLE)

GERMAN 3·78

FRENCH 1·03

ROMANSCH ·05

ITALIAN ·54

TOWNS WITH OVER 10 000 POPLN.

Jura

Plateau

Alps

VILLAGES
● MAJOR
· MINOR

Glaciers & Snowfields   BREITHORN

BLATTEN   R. LONZA   LANG GLACIER

N

0   km   3

Glaciers & Snowfields

Glaciers & Snowfields   Ridge

LÖTCHEN VALLEY: Distribution of Villages

(i) Using a marked vertical scale draw a cross-section diagram along the line XY on Map A. Add the following labels: *Jura Mountains*; *Swiss Plateau*; *Bernese Oberland*; *Rhône Valley*; *Pennine Alps*.

(ii) In which country and sea area does the water of each of the following rivers enter the sea:—*Rhine? Rhône? Ticino?*

(iii) (a) Name each of the lakes numbered 1–8. (b) What does the shape and location of these lakes suggest about their probable origin? (See p. 11, if necessary.)

(iv) Which region of Switzerland has the greatest proportion of arable land?

(v) (a) What links can you see between the distribution of population and the relief? (b) Suggest reasons for this distribution.

(vi) (a) What proportion of villages in the Lötschen Valley lie north of the river? (b) What explanation can you offer for this settlement pattern? (Hint: aspect.)

(vii) (a) What proportion of the total population lives in the towns listed at left? (b) What does this suggest about the importance of industry in the Swiss economy?

(viii) What proportion of the towns over 10 000 inhabitants are in (a) the Jura; (b) the Mittelland Plateau and (c) the Alps?

(ix) What proportion of the Swiss population speak (a) German; (b) French; (c) Italian and (d) Romansch?

(x) What language would you expect to hear the inhabitants speaking in each of the following places: Zurich, Martigny, Geneva, Lugano, Bern?

(xi) What proportion of the land surface of Switzerland is (a) forested; (b) pasture; (c) arable land; (d) waste?

(xii) Which famous tunnel(s) would one use on a railway journey (a) from Basel to Milan; (b) from Bern to Milan; (c) from Geneva to Milan? (*Map p. 140*).

ARABLE	PASTURE	FOREST	WASTE e.g. snowfields glaciers bare rock lakes

LAND–USE IN SWITZERLAND

THE SWISS PLATEAU is the most important region in the country, containing . . .

about two-thirds of the entire Swiss population;
the most extensive areas of cultivable lands;
all the large and most of the smaller industrial towns;
the Federal capital (Bern) and the country's major river port (Basel).

Although usually called a plateau, this region is not in fact particularly flat except at the base of the Jura. The photograph shows a typical landscape near Lausanne: notice the patch-work of small fields (mostly under pasture) and the hill-top site of the small town. (*Can you spot the old town walls?*) The underlying rocks are mainly sandstones, but their surface lies buried beneath extensive sheets of glacial sands and clays dumped there by Alpine ice-sheets and glaciers during the Ice Ages.

Farming on the Plateau. The better soils are extensively cultivated, mainly with fodder crops such as oats, hay and lucerne, for mixed farms are the rule and cattle are the mainstay of most farmers. Many dairy animals are reared, and the Swiss have a reputation for producing high-quality cheese, butter, tinned cream and condensed milk, as well as milk chocolate. Most farmers also keep pigs, fed mainly on skimmed milk returned from the dairies after the cream has been removed. Pork products, especially highly spiced sausages, form an important part of Swiss diet, and bread, cheese and sausage is the characteristic food of Alpine climbers. Orchard fruit, mostly apples, pears, plums and cherries, are grown in all parts of the Plateau, notably on sunny, south-facing hill-slopes. Areas of specialised crop production include (i) the south-facing slopes on the north shore of Lake Geneva, famous for their vineyards (*II: 67B*) and (ii) the Cantons (*political divisions*) of Vaud, Bern and Zurich (*map p. 134*), where market-gardening is important. ((*6*) *Suggest why?*)

Industry on the Plateau employs about 4 in every 10 of Switzerland's working population. All towns shown on the map on page 135 have factories of some kind, though the main industrial centres are Zurich, Basel, Bern and Geneva. A great variety of manufactured goods is produced, but engineering, watch-making, textile, chemical and food-processing industries predominate. The growth of these industries is remarkable in view of the fact that Switzerland has virtually no coal, few minerals, no direct access to the sea, a small home market and a mountainous terrain. To overcome these disadvantages the Swiss have concentrated on producing very high-quality articles of fairly small bulk but great value, e.g. machine tools, watches and optical instruments. (7) In what way does this favour a country which needs to import most of its raw materials and sell its manufactured products in distant foreign markets? Some idea of the fine precision of Swiss engineering and of the lengths to which Swiss manufacturers go in their attempts to satisfy the particular needs of customers can be obtained from the following:—

"The tiny machine that is a Swiss watch can be reduced to such a degree that it will fit into the setting of a woman's ring. Yet even at this size, it has more than seventy different parts, some of them so minuscule that they are nearly invisible. . . . Our factories make many different types of watches for many varied purposes. . . . Braille watches for the blind with a hinged cover or crystal to protect the dial and reinforced hands. . . . medical watches to assist doctors and nurses. All these have special features for pulse-taking, for timing injections of local anaesthesia or the development of X-rays. Sports car watches which permit rally and road race drivers to clock precisely their speeds and distances and help them in calculations to obtain maximum performance. Yachting watches . . . rowing watches . . . fishing watches. . . . There are also golf counters; footage timers for the professional cameraman; telephone watches to cut down on telephone bills; wrist navigators for airline pilots; slide-rule watches for engineers; parking alarm watches to warn you when to move your car; all sorts of sport timers; and watches to warn the skin-diver when he should start his ascent and telling him how to time it. I could go on. There are many others, but my point is made. Whatever your profession or hobby is, there is a watch for you."*

* Dr. J. J. Bolli in *The Guardian*.

Details of the main Swiss industrial towns are given below. Their factories use hydro-electric power generated by a large number of stations located both on the Plateau and in the Alps (*map p. 135*). Apart from the traditional skills and enterprise of her people, h.e.p. is the country's only important industrial asset.* The principal power-stations on the Plateau are on the Rhine between Schaffhausen and Basel, and 30% of the country's installed power is produced in the Cantons of Zurich and Basel.

* Virtually all Swiss hydro-electricity power sources are now tapped. By 1975 the thermal and nuclear proportion of Swiss electricity generating capacity will rise to more than 25%.

CHIEF INDUSTRIAL TOWNS OF SWITZERLAND

ZURICH is by far the largest Swiss town. It grew up where the main east–west route on the Plateau crosses the River Limmat on its exit from Lake Zurich. Its position makes it a natural focal point and market centre for the Eastern Plateau. During the past century the development of large local supplies of h.e.p. has led to a great expansion of Zurich's industries. The chief of these are textiles (mainly silk and cotton) and electrical and mechanical engineering (e.g. turbines, dynamos, locomotives). Zurich is also the main Swiss commercial and banking centre and has a world-famous Institute of Technology.

BASEL is an ancient Roman city, situated at the convergence of three major international routeways:—(i) from Rotterdam via the Rhine Valley; (ii) from Northern Italy via the Alpine tunnels and passes and (iii) from Paris, via Dijon and the Belfort Gap. *(8) With the help of an atlas draw a sketch-map to show these routes.* Basel is a large river-port handling nearly one-half of Switzerland's foreign trade. Its three miles of quays can be reached by 2000-tonne Rhine barges. The docks include a 'free port', or bonded warehouse, which contributes extensively to Switzerland's commerce. The annual Swiss International Fair is also held in Basel: this event lasts 10 days and attracts some 700,000 buyers, including foreigners from 70 countries. The city also has textile, engineering and chemical industries, the latter giving employment to one-third of Basel's working population. Raw materials (e.g. coal, coal-tars and petroleum) are obtained by barge from the Ruhr. Thousands of tourists arriving annually in Switzerland by road and rail enter the country at Basel, and the money they spend adds to the prosperity of this wealthy city.

GENEVA is also an important entry-point to Switzerland. *(9) Explain why, after studying its geographical position in an atlas.* The city is best known as a centre for international meetings and conferences —a function derived from Switzerland's long history of political neutrality. It is the headquarters of the International Red Cross and several U.N. Agencies. Geneva also has important industries, especially jewellery- and watch-making, chemicals (notably perfumes) and engineering. Many Swiss banks and commercial firms have their main offices in Geneva.

BERN is the Federal Capital of Switzerland. *(10) How does its geographical position favour this function? (Use an atlas.)* In addition to its many government buildings and offices the city has important engineering and watch-making industries.

Other smaller, but important industrial towns include *Winterthur* (electrical equipment, e.g. turbines and generators); *St. Gallen* (cotton textiles and embroidery); and *Neuchâtel, Le Locle* and *La Chaux de Fonds* (watch-making). In addition there are many industrial firms in remote valleys, using local supplies of h.e.p. Such concerns include small family-owned workshops making watches and large chemical works such as aluminium refineries.

THE JURA are a series of plateaus and simple fold ranges, rising to over 1700 m in the south, which mark the frontier between France and Switzerland. Although composed mainly of limestone they have a heavy rainfall (875–1125 mm) and so are well wooded, and there are extensive upland pastures. Cattle rearing forms the basis of Jura farming, the animals being taken to the uplands to graze from May until September, and brought back to sheltered valley farms for the winter. ((11) *What is this practice called?*) There are many food-processing factories in the valleys, and the region is well known for its Gruyère cheese, condensed milk and milk chocolate. The forests are another source of wealth, the timber being used for fuel, for building purposes and to make furniture and wood-carvings. Another traditional Jura occupation is clock- and watch-making; many villagers make watch-parts at home during the winter, but large-scale factory production is concentrated in the larger towns (*see notes opposite*). Communications in the Jura are difficult because of the steepsided, parallel ranges. Roads and railways keep to the longitudinal valleys (*vals*), and cut through the ridges by way of deep limestone gorges known as *cluses*.

THE SWISS ALPS form part of the great series of fold mountain ranges described on page 10. When approached from the Plateau they rear up abruptly to such prodigious heights as to appear impenetrable. Yet in fact they are crossed fairly easily, and well-trodden trans-Alpine routes have linked the Mediterranean lands to North-west Europe for over 2000 years. The 'grain' of the Alps runs N.E.–S.W., and is followed by many deep, glaciated, longitudinal valleys. The greatest of these troughs (followed in their upper courses by both the Rhine and the Rhône) divides the Swiss Alps into two main parts and provides a great natural highway through their very heart (*maps p.* 134 *and overleaf*). Tributary *transverse* valleys (so called because they cut across the grain) enter the Rhine–Rhône trough at right angles and are being cut back deeply into the adjoining ridges of the Bernese Oberland and Pennine Alps. These mountains are also being dissected by rivers draining northwards to the Plateau and southwards to the North Italian Plain, and where the heads of transverse valleys approach one another from opposite sides of a ridge it is comparatively easy in summer to walk over the crest through a *col* or pass (*see diagram overleaf*). Large areas of the Swiss Alps are completely uninhabited.

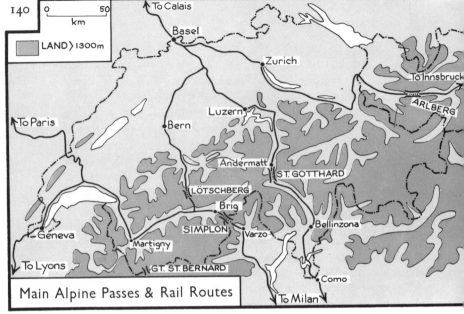

To Calais
Basel
Zurich
To Innsbruck
ARLBERG
To Paris
Luzern
Bern
Andermatt
ST. GOTTHARD
LÖTSCHBERG
Brig
Bellinzona
SIMPLON
Varzo
Geneva
Martigny
To Lyons
GT. ST. BERNARD
Como
To Milan

0   50
km

LAND > 1300m

Main Alpine Passes & Rail Routes

There are, for example, 1820 square kilometres of permanent snowfields and glaciers, and much of the higher land which is free from snow in summer is too steep, rocky, cold, wet and infertile to support a farming community. Most inhabitants live in the deeply entrenched glaciated valleys, where the climate is milder and there are some boulder-clay and alluvial soils. (*12*) Describe the distribution of the settlements in the diagram below. (*13*) Why are there no settlements on one side of the valley? (*Hint: sunshine.*) (*14*) Suggest three reasons why village X was built on that particular site. (*15*) What danger annually threatens such Alpine villages during the spring thaw?

Farmers in the Alps rely chiefly on dairy cattle for their liveli-hood. As in the Jura, the animals are taken up to graze on mountain pastures during the summer and brought back to valley

South
North
Transverse Valley
Road

TYPICAL SETTLEMENT PATTERN OF AN ALPINE VALLEY

farms for stall-feeding in winter. Milk is sent daily in churns, tanker-lorries and polythene pipe-lines (from remote summer pastures) to co-operative dairies in the valleys. Some of the latter are famous for particular varieties of cheese, e.g. Emmental. Fragmentary patches of arable land are cultivated, mainly with fodder crops such as barley, roots and hay, but orchard fruit and vines are grown on specially favoured south-facing slopes. When possible, water from mountain torrents is led along wooden troughs for irrigation, especially in the Ticino region where summers are hot and dry.

The steeper, lower Alpine slopes are thickly forested, mainly with conifers, and timber is widely used for fuel, for building châlets and in wood-working industries. Great care has to be taken, however, not to remove too many trees, for they help to protect the villages from disastrous avalanches. ((*16*) *How?*)

Tourism. The past half-century has seen a phenomenal growth in foreign travel and tourism. This has been due largely to (i) improvements in road, rail and air transport; (ii) a general increase in living standards in Europe and North America and (iii) the granting of annual paid holidays. No country has benefited more from these developments than Switzerland. Her magnificent mountain and lake scenery attracts no fewer than $1\frac{1}{2}$ million foreign visitors every year, and the money they spend in Switzerland represents about 3% of the total national income. The main scenic attractions are naturally in the high Alps, and a remarkable system of mountain roads and railways, together with ski-lifts, hotels, shops, restaurants and kiosks in nearly every main valley, cater for the tourist's every need. Throughout the Swiss holiday industry the emphasis is on individual service . . .

". . . the (hotel) industry consists predominantly of small and medium-sized enterprises, most of them kept within the family so that our guests can rely on personal attention, which has always been a characteristic of Swiss hotels . . . both resorts and hotels have been modernised in order to offer our visitors the highest degree of comfort. . . . Swiss hotel-keepers keep in mind the elementary fact that the way to a guest's heart is through his stomach. They have therefore made great efforts to improve the quality of cooking, taking in account changes in eating habits, so that the culinary arts have reached a high level."*

Famous Swiss resorts include: St. Moritz, Luzern, Lugano, Locarno, Zermatt, Grindelwald, Interlaken. (*17*) With the help of an atlas sort these into three groups, as follows:—(i) High Alps (winter sports); (ii) northern edge of Alps; (iii) Ticino valleys and lakes (south-facing and sunny).

* Dr. A. Pfister, writing in *The Guardian.*

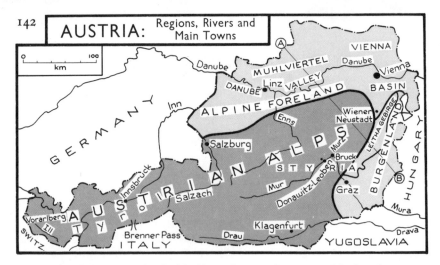

AUSTRIA: Regions, Rivers and Main Towns

AUSTRIA may be divided into three main geographical regions, as shown above. Notice that the cross-section below is drawn along line AB on the map, i.e. it represents the compact eastern portion of the country. (*18*) Which of the three regions extends westwards to include the elongated part of Austria known as Vorarlberg and

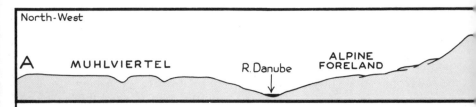

North-West

A MUHLVIERTEL   R. Danube   ALPINE FORELAND

THE DANUBE LANDS TO THE NORTH

These consist of (i) the flat-topped, dissected plateau of Muhlviertel; (ii) the Alpine Foreland; (iii) the Danube Valley. *Muhlviertel*, on account of its raw winter climate and thin, infertile soils, remains largely forested and scantily peopled. *The Alpine Foreland*, in contrast, consists of pleasant, gently rolling countryside, with occasional wooded sandstone ridges overlooking rich valley farmland. The latter is devoted mainly to cattle-rearing, though fodder crops are grown on patches of fertile boulder-clay and loess. *The Danube Valley* in Austria varies in width from narrow gorges to broad basins across which the river follows a **braided** course between low gravel islands. The main area of lowland is the Vienna Basin, which is covered with fertile soils and contains the only extensive area of arable land in Austria. Here, too, are the largest and most productive farms, growing wheat, oats, maize and potatoes in rotation, as

well as vines and fruit on south-facing slopes.

The Danube can be a dangerous river to navigate, especially during the spring floods and in winter, when ice-floes are a constant hazard. Even so, for centuries it has carried trade between Central and Eastern Europe, and many Austrian towns are strung out along its banks. The most important of these are Linz and Vienna, details of which are given later.

THE ALPINE AREA

This makes up 70% of the country and consists of a series of parallel-fold mountain ranges, separated by deep glacial valleys. ((*19*) *Name three of the latter.*) Towards the east the ranges gradually fan out and the basins between them become wider and lower. These basins, e.g. at Klagenfurt and Graz, contain fertile soils and are intensively cultivated. In contrast to the pastoral farmers of the Swiss Alps most Austrian peasants practise

Tyrol? (*20*) From the map and an atlas suggest, with reasons, the directions in which travel in Austria is (*a*) relatively easy and (*b*) relatively difficult. (*21*) Into which major river do Austrian rivers drain? Can you see any exceptions?

Industry in Austria. Iron and steel manufacture has been carried on in Austria on a small scale for centuries. The iron ore which mostly comes from the Erzberg mines in Styria, was originally smelted by local charcoal, but now coke is used, imported from the Ruhr and Silesia. The main centre is Donawitz-Leoben, from which an industrial conurbation now extends along the Mur-Murz valley to Bruck (*see map*). The old-established steel mills at Linz have also been expanded in recent years. The Austrian steel industry, though relatively small, maintains a foothold in foreign markets by producing very high quality goods, some 700 different first-grade steels being exported all over the world. There is also a steady export of engineering products such as oil-drilling equipment, optical and precision instruments and special steel-making plant.

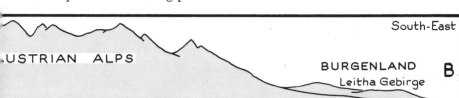

South-East

AUSTRIAN ALPS

BURGENLAND **B**

Leitha Gebirge

polyculture, i.e. they grow a great variety of crops, e.g. maize, wheat, fruit, tobacco and vines. Most farm-holdings, however, are extremely small (between 0·8 and 4·0 ha), and in the remote higher valleys much of the arable land is rugged, steep and infertile. Traditional hand methods of tillage are still employed and crop yields are low. The upper limit of cultivation lies between 1350m and 2000m, according to the aspect and steepness of particular valleys. In Vorarlberg only 1%, and in the Tyrol only 3% of the land is cultivable: here livestock farming, especially dairying predominates. Much farm work in the Austrian Alps is seasonal, for winters are long and cold and there is heavy snowfall.

Alpine Austria also has a variety of old-established industries. These include (i) the manufacture of embroidery and textiles in small towns along the Swiss border; (ii) the mining of iron and non-ferrous ores and salt in Styria and Salzburg and (iii) the production of timber for fuel, construction, resin and turpentine: much of Alpine Austria is forested with conifers. In recent years an ancient iron and steel industry in Styria has been greatly expanded, and the production of h.e.p. is also rising. Details of these modern economic developments are given above and overleaf.

THE EASTERN LOWLANDS

Between the Alpine foothills and the Hungarian border lies a monotonous, gravel-strewn plain called the Burgenland. In spite of its bitterly cold continental winters and scanty rainfall (less than 20") Burgenland contains the highest proportion of improved farmland (67%) of any Austrian province. Water is raised from below ground by 'beam and bucket' wells, and the steppe-like plain is largely given over to stock-rearing and extensive wheat cultivation. A low ridge (the Leitha-Gebirge) protrudes into the plain, and here much orchard fruit and grapes are grown on south-facing terraces.

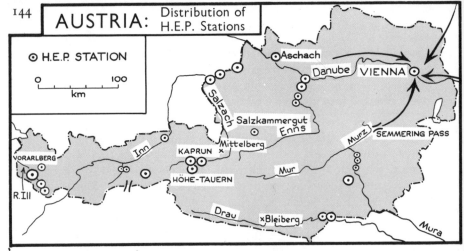

AUSTRIA: Distribution of H.E.P. Stations

⊙ H.E.P. STATION

0 ——— 100
km

The main drawback to these developments is Austria's lack of coking-coal. On the other hand there is a small coalfield, a petroleum and natural gas field and several deposits of lignite (*map p. 133*). Austria's main internal source of power, however, is hydro-electricity. Details of present production are given below. A greatly expanded output of electricity is planned for use in the production of alloy steels and the refining of **non-ferrous** metals. A variety of these are mined in Austria, including zinc, copper and lead in the Mittelberg, lead and zinc at Bleiberg and bauxite in Vorarlberg. Besides these developments in heavy industry and metallurgy, every effort is being made to expand such industries as *light engineering, textiles, chemicals and food-processing.*

Hydro-electricity supplies more than 80% of the power used in Austria, and in addition is exported in considerable and growing quantities to West Germany and Czechoslovakia. Most of the power stations lie in the drainage basins of the Rivers Enns, Salzach, Inn and Ill. (*See map:* (*22*) which of these rivers does *not* drain to the Danube?) Large new power stations have been constructed or enlarged during the past twenty years. The most notable schemes are those at Kaprun-Hohe Tauern and at Aschach on the Danube: the latter is Europe's largest h.e.p. station outside the U.S.S.R., producing 10% of all Austrian hydro-electricity. Other important projects are due for completion shortly. (*23*) What was the percentage increase in h.e.p. production in Austria between 1937 and 1970?

H.e.p. is fed into the Austrian 'grid' and distributed throughout the country for domestic and industrial uses. It is also the

H.E.P. Production in Austria	
1937	2800 m. kWh.
1970	18 185 m. kWh.

main source of power on the railways and:—

"has helped to bring about a significant rise in rural living standards, mainly through stimulation of the tourist trade. Standards of accommodation have been improved and such facilities as ski-lifts installed." (P. J. M. Bailey, *Geography*.)

Although industry in Austria is expanding, it is likely that there will always be a surplus of h.e.p. for export; thus a new super-grid (i.e. very high voltage) scheme has been constructed to serve a wide market in Central Europe.

Vienna stands on a broad terrace above the right bank of the Danube, where the river breaks out from its mountain course. Thus the city occupies a highly important strategic position, guarding the approaches to the Rhineland from South-east Europe. In medieval times Vienna was for centuries a Germanic and Christian bastion against the onslaughts of Moslem Turks. It also became a great trading centre, for it lay at the focus of the following important routes: (i) from Poland and the Baltic via the Moravian Gate; (ii) from Venice and Trieste via the Semmering Pass; (iii) from the Rhineland via the upper Danube; (iv) from Eastern Europe and the Orient via the lower Danube. Each of these routes is indicated by an arrow on the map opposite.

The Opera House, Vienna—the city's many ornate buildings are a legacy of its former Imperial grandeur (see overleaf).

(24) Copy the relevant part of the map and, with the help of an altas, add appropriate labels to each of the arrows.

Main industries of VIENNA
Food processing
Engineering
Printing
Tailoring
Manufacture of:—
clothing
gloves
silks
electrical
apparatus

Until 1919 Vienna was the capital of the Austro-Hungarian Empire and a major centre of European politics, culture and art: today it retains some international importance, for United Nations agencies on Atomic Energy and Industrial Development are located there. It also remains the capital of Austria and its 1·64 m. inhabitants form more than 25% of the country's total population. How to feed such a large urban population is one of Austria's main economic problems, for although the city lies adjacent to the fertile plains of the Vienna Basin and Burgenland, large food imports are still necessary. Most of Vienna's population work in the city's varied industries, some of which are listed here. (25) Which of these industries reflects Vienna's former importance as a leader of fashion?

Vienna also attracts large numbers of foreign tourists, who visit the city's splendid imperial buildings, boulevards, theatres and art collections. Formerly Vienna was an international banking centre, but this function declined following the Nazi persecution of the city's Jews, who included many prominent bankers.

OTHER CITIES OF AUSTRIA

Graz, with a population of 255 000, is Austria's second largest city. It originated as a fortified town, commanding the approaches to the upper Mur valley, and for centuries has been a market centre where Alpine and lowland produce is exchanged. It is also a cathedral town and has an ancient university. Many old-established industries, e.g. textiles, clothing, leather and brewing, are located in the town. To these have recently been added mechanical engineering, vehicles and chemicals, stimulated by the availability of hydro-electric power from the adjoining Alps.

Linz (population 210 000) originated as a Roman defensive site. For centuries its chief importance was as a bridging point of the Danube, carrying the ancient 'salt road' from Salzburg northwards to Bohemia. Today it is the administration centre of upper Austria, a busy river-port and a rapidly expanding industrial town. Its principal industries are steel and chemicals. Iron ore from Styria and Sweden is smelted with Ruhr coke, and newly constructed electric-arc furnaces produce high-quality steels. The chemical works use hydro-electricity to produce nitrates, fertilisers and sulphuric acid. Other industries in the vicinity of Linz include engineering, vehicles, glass-making and textiles.

Innsbruck and Salzburg are best known as tourist centres for the Tyrol and the Salzkammergut respectively. The towns are also similar in several other respects: both are beautifully situated with magnificent views of the Austrian Alps; both control important north–south routeways ((26) explain, after referring to the map on page 142 and an atlas); and both use hydro-electricity in a variety of industries. These include engineering and chemicals in both towns, and aluminium refining in Salzburg, using bauxite imported from Yugoslavia.

CHAPTER 7

Southern Europe: (i) Italy

A RELIEF map of Italy shows several varied and distinct regions —great Alpine ranges in the north; flat, monotonous plains in the Po Basin; a long, narrow, mountainous and often barren peninsula; and the mountainous and typically Mediterranean islands of Sardinia and Sicily. For centuries these territories were divided into a maze of small independent kingdoms and principalities, and it was not until 1860 that the unified nation-state of Italy came into existence. Since unification, moreover, the development of modern Italy has been bedevilled by the fact that economic progress has been mainly confined to the northern half of the country.

"Each morning in the Milan railway station a poignant scene demonstrates the extent to which Italy, for all its seeming unity, is still a nation of deep and tragic divisions. On one track, first-class sleeping cars, just in from Switzerland's Gotthard Pass, are unloading night-travelling Milanese returning from business trips to the commercial and banking cities of northern Europe—Zurich, Brussels, London and Amsterdam. The self-assured international businessmen, shaved and fresh in blue or grey pin-striped suits and accompanied by chic women in furs and tweeds, walk briskly behind porters.

"On another track, at the same moment, ill-smelling coaches from Italy's South are letting out travel-weary Sicilians and Calabrians. Short, dark men wearing green moleskin suits peer worriedly about the station, looking for a relative or an old friend. The women, unkempt and frightened, wearing black shawls and skirts and clutching cloth bundles and cardboard suitcases, huddle in protective coveys around their thin, grave children, who are absorbing with wide eyes the bustle and confusion.

"The passing Milanese businessmen cast distasteful glances at these intruders, whom they call *terruni*, from the Italian *terra*, or earth. The invaders average 115 a day, and some Milanese, who have only two or three children to a family, fear that the prolific *terruni* may some day breed them out of existence. Already Sicilians control the city's fruit and vegetable markets and Neapolitans the retail textile business. The city is ringed with a shanty-town belt of 200,000 of the newcomers.

"Where is the border that the *terruni* cross on their flight from the south? It is roughly the boundary between the old Bourbon kingdom to the south and the Papal States to the north. It begins near Rome and runs north-east to the Adriatic below Ancona."*

This contrast between the two 'Italys'—the thriving, bustling,

* Herbert Kuhly, *Italy* (*Sunday Times* World Library).

prosperous North and the backward, lethargic and poverty-stricken South—can be explained partly by their history but even more by their geography. (*1*) How far is the contrast reflected in these maps and statistics?

Later in this chapter we shall see that the Italian Government is making determined efforts to improve the lot of its southern citizens, but the figures in the table opposite do not encourage optimism. One must bear these sombre points in mind when reading about 'The Italian Miracle', i.e. the extraordinary way in which Northern Italy managed to recover from the devastation and economic collapse of the Second World War to become a vital, go-ahead centre of European industry, commerce and the arts,

ITALY—the 'North': Milan shopping arcades.

ITALY: Relief, Regions and Distribution of Larger Towns

	The 'North'	The 'South'
Population	34 140 000	18 961 000
Total births p.a.	c 209 000	c 264 000
Agricultural workers	2 057 000	1 966 000
Cultivable land	172 000 km²	104 000 km²
Employed in manufacturing	6 219 000	1 829 000
Employed in service trades	4 787 000	2 013 000
Unemployed	175 000	133 000
Looking for a first job	182 000	173 000
Average per capita income	£780	£381*

* Wages in the South vary enormously. In Calabria they average £211, while in parts of southern Sicily they are below £100 p.a.

Mean Rainfall (mm)	J	F	M	A	M	J	J	A	S	O	N	D	Total
Milan	60	58	68	85	103	83	70	80	88	118	108	75	996
Palermo	98	83	70	65	33	15	8	15	38	98	98	113	734

ITALY—the 'South': Sicilian village street.

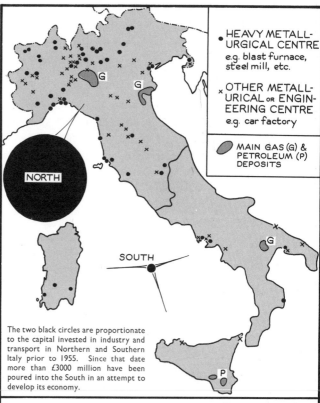

- HEAVY METALL-URGICAL CENTRE e.g. blast furnace, steel mill, etc.

× OTHER METALL-URICAL or ENGIN-EERING CENTRE e.g. car factory

MAIN GAS (G) & PETROLEUM (P) DEPOSITS

NORTH

SOUTH

The two black circles are proportionate to the capital invested in industry and transport in Northern and Southern Italy prior to 1955. Since that date more than £3000 million have been poured into the South in an attempt to develop its economy.

ITALY: METALLURGICAL & ENGINEERING CENTRES

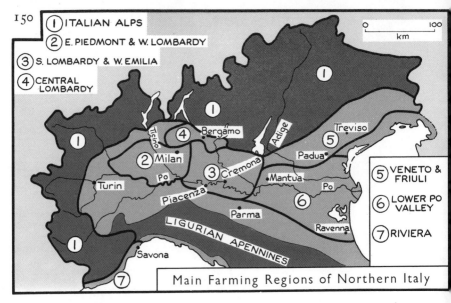

① ITALIAN ALPS

② E. PIEDMONT & W. LOMBARDY

③ S. LOMBARDY & W. EMILIA

④ CENTRAL LOMBARDY

⑤ VENETO & FRIULI

⑥ LOWER PO VALLEY

⑦ RIVIERA

Main Farming Regions of Northern Italy

NORTHERN ITALY includes several contrasting geographical regions (*map p. 148*) but has at its heart the fertile plains of the Po Basin. Ever since Roman times the agricultural wealth of these plains has given Northern Italy an advantage over the South.

They contain most of Italy's best arable land, and form the only part of the country to be blessed with a plentiful and well-distributed rainfall. (*From the figures on page 149 (2) state the total mean annual rainfall in* (a) *Milan and* (b) *Palermo.* (3) *Describe and explain the contrasts in the seasonal distribution of rainfall in each of these places.* (*If necessary see pp. 14–18.*)) Other geographical advantages which favour Northern Italy include (i) relative nearness to Western and Central Europe, to which it has been linked by trans-Alpine trade routes since the early Middle Ages; (ii) easy access to such old-established ports as Genoa and Leghorn; (iii) large supplies of h.e.p. from the Alps and Northern Apennines; (iv) recently-discovered supplies of natural gas (*p. 149*); (v) the patient effort and skill of its people.

Agriculture. Prior to the Ice Age the Po Basin was occupied by a broad inland sea—an extension of the Adriatic. During the past million years this shallow gulf has gradually been infilled with alluvium and rock fragments swept down from the Alps and the

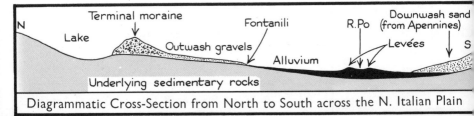

Diagrammatic Cross-Section from North to South across the N. Italian Plain

Northern Apennines. The diagram shows the location and origins of the main types of deposit.

Soils on these deposits vary greatly in fertility and this, together with variations in water supply and climate, accounts for the diversity of farming which characterises Northern Italy. The main farming regions are shown in the map opposite and described in the notes below. (4) Use the notes to make a *large* labelled map entitled 'Farming in Northern Italy'.

MAIN FARMING REGIONS OF NORTHERN ITALY

The Italian Alps are trenched by many deep, sheltered glaciated valleys where arable farming is practised on south-facing slopes up to about 1500 m. Hardy cereals and hay are grown on the higher slopes, and maize, hay and vines on lower ground. Around the shores of the many 'ribbon' lakes the winter and spring climate is noticeably milder ((5) *Why?*) and many lakeside farmers grow orchard fruit, e.g. peaches, apricots and olives, and a variety of vegetables (notably tomatoes, which are canned and exported). In the higher valleys pastoral farming predominates; the cattle and sheep are taken up to high alpine pastures during the summer.

Eastern Piedmont and Western Lombardy are best known for their large, mechanised farms devoted mainly to irrigated rice cultivation. The area under rice grew rapidly during the mid-19th century, and Northern Italy is now one of the principal rice-growing districts outside S.E. Asia. The rice is hand-planted by girls recruited annually from nearby towns, but is harrowed and harvested with machines. To maintain soil fertility the rice is grown in rotation with other cereals, and much artificial fertiliser is added.

Southern Lombardy and Western Emilia form the only important dairy-farming region in Italy. The land is low-lying, with rich alluvial soils, and farmers obtain as many as seven crops of hay in one summer. The meadows are irrigated with river water and also by warm-water springs (*fontanili*). ((6) Locate the *fontanili* on the diagram and explain their formation.)

Co-operative dairies in such towns as Cremona, Parma and Piacenza bottle most of the milk for sale in the great industrial cities of the Plain. Large quantities of butter and cheese are also made. Cheese forms part of every Italian meal, and the traditional cheeses of Northern Italy (notably Gorgonzola and Parmesan) are world-famous. Skimmed milk for the dairies is returned to the farmers, who mix it with maize and feed it to their pigs. Pork in various forms (e.g. *Bologna sausage*) is a traditional product.

Central Lombardy has long been renowned for its remarkable fertility and **polyculture** (i.e. variety of crops). "Such is the fertility of this country", wrote the English traveller Thomas Coryate in 1608, "that I thinke no Region or Province under the Sunne may compare with it. . . . For as Italy is the garden of the world, so is Lombardy the garden of Italy. . . . The first view thereof did even refolliate my spirits and tickle my senses with inward joy." Today the region still produces high yields of maize (for *polenta*, a staple food of the poorer peasants), clover, wheat (for bread, macaroni and spaghetti), hay, hemp, vegetables, orchard fruit and vines, and many dairy cattle are reared.

Veneto and Friuli also produce a variety of crops, though the emphasis is on wheat, maize and clover. Farms here are very small and the soils only moderately fertile, so that much manure and chemical fertiliser is required.

The Lower Po Flood Plain and Delta consist of drained alluvial soils; the main crops are rice, hemp and sugar-beet. In the delta the Government runs several 'Colonisation Centres', (i.e. large State Farms) where poverty-stricken peasants from elsewhere can begin a new life.

The Riviera. From Genoa westwards, as on the French Riviera (p. 74) flowers, fruit and early vegetables are intensively cultivated, often under glass, wherever the land on the sheltered, south-facing coast is flat enough to permit it.

Industry has been established in Northern Italy for centuries, but only during the past 20 years has the Po Basin emerged as one of Europe's great manufacturing zones. Lack of home-produced fuels made Italy a late starter in the Industrial Revolution, but from 1900 onwards the Alpine and Northern Apennine torrents were harnessed for hydro-electric power—power which gave the Plain's scattered textile, metallurgical and engineering industries a tremendous boost. After the Second World War the Italian Government deliberately encouraged a great industrial expansion, hoping that the export of manufactured goods would help improve the Italians' comparatively low standard of living.

> "The ebullient Italians, with their sturdy national virtues of imagination, humour and willingness to work, had the country functioning again very shortly after the war. There followed a period of astonishing reconstruction, generously helped by American aid. Today the North is the seat of Italy's amazing post-war boom . . . this prosperous region holds the key to the country's progress and new-found economic wealth. . . . Of the country's 10 major cities, all but Naples and Palermo are in the North. The economic flowering centres on a triumvirate of northern cities: manufacturing Turin, banking and shipbuilding Genoa and commercial Milan. Milan is the leader. Milan has supermarkets, corner filling stations, hire-purchase and a rising skyline of steel and glass office buildings, banks and apartments. As a Milanese once said, 'Rome has politics, ruins and the Pope, but Milan is the real capital—financial, commercial, industrial, musical, artistic, theatrical, publishing, jazz. . . . What more do you want?'"*

By good fortune large quantities of natural gas (methane) were discovered in Northern Italy in 1945, i.e. just when extra power was needed. The main gas-fields lie along the southern margin of the Po Basin (*map p. 149*) and an elaborate 'grid' of pipelines now delivers methane to factories and homes in all the chief cities

* Kuhly, *op. cit.*

Italy: Sources of Energy		
All Units in Tonnes, Coal Equivalent		
Petroleum	74 490 000	(95% imported)
Hydro-electricity	26 244 000	
Thermal-electricity	19 200 000	
Coal	13 950 000	(87% imported)
Geothermal-electricity	1 500 000	
Nuclear-electricity	1 440 000	

Part of the vast Fiat c factory in Turin. Note (i) t snow-capped Alps, (ii) t R. Po and its flood plain, (i the various electricity pylon

of the Plain. The gas is used as fuel for power-stations, factory boilers, domestic cookers and road transport; and as raw material for the manufacture of chemicals, e.g. nitrates and synthetic rubber. The growing importance of natural gas should, however, not obscure the fact that Italy still obtains her main supplies of energy from imported petroleum. (*7*) Draw bar diagrams to illustrate the figures given in the table. The **geothermal power** mentioned in the table is produced in the Larderello district of Tuscany (*map p. 162*) where steam of volcanic origin and under great pressure blasts its way to the ground surface through a number of vents (*soffioni*). Since 1900 Italian engineers have pioneered methods of harnessing this type of steam to generate electricity. (*8*) What proportion of the Italian electricity output comes from geothermal power stations? It is probable that this unusual source of power will become increasingly important in Italy.

Camillo Folchi is operating an electronic control panel to store new car bodies in the vast Fiat car factory in Turin. Half a million vehicles are turned out every year at factories in Turin (*Fiat* and *Lancia*) and in Milan and Brescia (*Alfa Romeo, O.M.* and *Autobianchi*). As a member of the Italian car industry, Camillo belongs to a well-paid, comfortably-off elite amongst Italian workers. With his weekly wage of £30, plus Christmas bonus and other benefits, Camillo earns on average 80% more than the Italian minimum-wage scale. He learned his trade at the Fiat apprentice school, plays the drums in a Fiat jazz group and races his cycle for the Fiat Sports Club. He is buying his transistor radio and motor-scooter on hire-purchase from Fiat, and lives in a new block of flats built by the company.

> "Look down from a balcony . . . and you will see the rolling green court of a company sports club where workers are playing *bocce*, or Italian bowls. Nearby is the company medical centre, the most modern of its kind in Italy, and near it is a company nursery. Beyond is the local church with its company-financed youth recreation centre. What you are looking at is a little Fiat world, equipped to fulfil a worker's every need."*

The spectacular post-war expansion of the Fiat company and the prosperity of its workers is one of the success stories of modern Italian industry. It is rivalled, however, by equally successful ventures in many other branches of engineering, e.g. the production of electrical equipment, agricultural machinery, ships, aircraft, sewing-machines and typewriters. Metal-working is now Italy's 'key' industry (a place held in former times by textiles) and the main metal used is steel. In recent years the Italian steel

* Kuhly, *op. cit.*

industry has grown at a rate faster than that of any other West European country, a direct result of Italy's membership of the 'European Coal and Steel Community' (E.C.S.C.) which she joined as a founder member in 1952. With few raw materials of her own, Italy now receives guaranteed regular supplies of scrap iron from other members of E.C.S.C., and imports of Ruhr coke at favourable low prices.

With an annual steel consumption of c. 21 m. tonnes, and a steady export trade in high quality steel goods, Italy is now one of the leading industrial countries of Western Europe. This success is marred, however, by some serious problems. Although new steel plants have recently been built at Taranto and Naples the great bulk of Italian steel and engineering works are still inside the 'triangle' Milan-Turin-Genoa. Since 1950 these 'boom' cities have attracted hundreds of thousands of unemployed labourers from rural areas, mainly in the South. Far more migrants have arrived than there are jobs to be filled, so the great northern cities face constant crises of unemployment and shortages of houses, schools, doctors, hospital beds, buses and other essential services. The problems are immense, for some 300 000 Italians leave the land every year. The urgent efforts to provide more urban jobs in the South are described on pages 170–173.

ITALY
Main Iron and Steel
Producing Centres
Towns with blast
furnaces are underlined

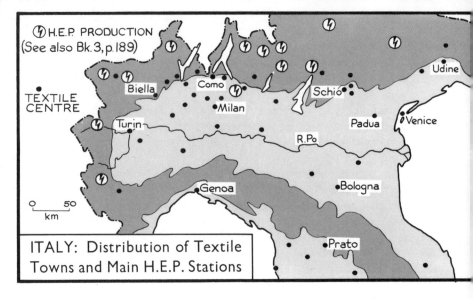

(See also Bk. 3, p. 189)

ITALY: Distribution of Textile Towns and Main H.E.P. Stations

Although the Turin–Milan–Genoa 'triangle' is now emerging as the major engineering region, other industries in Northern Italy remain remarkably scattered. This dispersal originated in the days of water-power, and has persisted because (i) Italy has no important coalfields and (ii) her industries depend very largely on electricity and natural gas as sources of power. The widespread distribution of one industry is clearly shown on this map—nearly every town of any size has some form of textile manufacturing. As in other European countries, the Italian textile industry has contracted during this century as a result of keen competition from such 'new' producers as Japan and India. Italian

---

OTHER IMPORTANT INDUSTRIES IN NORTHERN ITALY

Food-processing is important in most of the towns of the North Italian Plain. The main products and centres are:

Cheese—at Bergamo, Parma, Lodi, Pavia, Cremona and Piacenza.
Meats, sausages, pastes and sauces—at Bologna and Milan.
Confectionery and pasta—at Milan.
Rice—at Vercelli.
Wines and liqueurs—at Modena, Asti and Turin.

(9) Show this information on a *large* labelled sketch-map.

(10) Find out which famous Italian foods or drinks are associated with the following words:—*Cinzano, Chianti, Gorgonzola, Risi e Bisi, Canelloni.*

Chemical manufacturing industries have expanded rapidly in Northern Italy since 1945. The chemical works are of two main kinds:—(i) those (mostly located in the Alpine foothills) which use h.e.p. to 'fix' nitrogen from the air and produce nitrates; (ii) those which use natural gas as a raw material to produce petrochemicals such as plastics, drugs, synthetic resins and perfumes. At Ravenna is the largest plant in Europe for the production of synthetic rubber.

Shipbuilding and marine engineering are carried on chiefly at Genoa and Trieste, and to a lesser extent at La Spezia and Leghorn. During the Second World War the Italian merchant fleet was virtually destroyed, but it has been rebuilt and expanded mainly by the output of ships from Italian yards. Vessels built include tankers, cargo-boats and luxury express liners for the N. Atlantic run. At present Italy has the world's fifth largest merchant fleet.

The liner Michelangelo *being launched in a Genoese shipyard.*

textiles shared, however, in the general post-war boom. Reasons for their recovery included the closing of less efficient mills, the re-equipment of others with modern machinery and a sharp rise in the production of man-made fibres. Another very important reason is that Italian fashions have been the rage in cities as far apart as Newcastle and New York, Paris and Pittsburg.

Labelling cheeses at a factory in Parma.

This photograph shows the fine *autostrada* (motorway) which leads north out of Milan in the direction of the St. Gotthard Pass. Like most of the modern routes shown on the map, the Milan–Como *autostrada* follows the path of an ancient Roman road. Two good reasons why these particular routes have been used consistently for some 2000 years are apparent from the map. (*11*) Write them down. (*Hints: relief; drainage.*) Most towns of the Plain are also very ancient, many having been occupied continuously since pre-Roman times. The towns can be sub-divided into three main categories,

(i) settlements at the foot of the Apennines (strung out along (*12*) which famous Roman road?);

(ii) settlements situated on firm ground at the edge of the Po flood plain;

(iii) settlements situated where routes across the Alps and Apennines open out on to the Plain.

(*13*) Into which of these categories do each of the following fall:—Milan, *Alessandria, Bergamo, Parma, Brescia, Verona Vicenza, Reggio, Modena, Bologna, Treviso, Turin?* There are also towns, e.g. Piacenza and Cremona, at bridging points of the River Po. Few towns are built on the banks of the *lower* Po, however, because of the danger of flooding. (*14*) How far from the sea is the lowest large riverside town? Settlements are sparse, too, along the Adriatic coast, for the land there is swampy and fringed by shallow lagoons that are difficult to reach from the open sea. (For the exceptional case of Venice see page 175.)

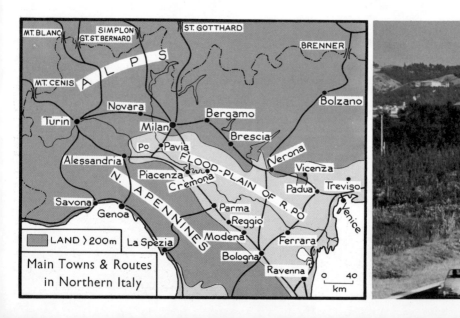

Main Towns & Routes in Northern Italy

LAND > 200m

MILAN (1 680 000) has been a major route centre since Roman times. It grew up where trans-Alpine Roman roads converged, and has always derived much of its importance from trade via the Alpine passes. The St. Gotthard and Simplon tunnels, opened in 1882 and 1906 respectively, made Milan the greatest centre of international railways in Italy. Now six *autostrade* also converge on the city. Milan dominates the whole Plain, and in this century has become the undisputed industrial, commercial and financial capital of Italy. "Excellent rail communications, hydro-electric power (and more recently gas), skilled and hard-working labour and enterprising management account for Milan's position as the most important manufacturing city in Italy. . . . In the centre of the city imposing blocks of offices, some of skyscraper dimensions, house the head offices of banks, insurance companies and Italy's most important industrial concerns; the Stock Exchange is the busiest in the country and the Milan Fair provides an international shop window for Italian industry." (D. S. Walker, *The Mediterranean Lands* (Methuen).)

Milan also has two universities, the leading publishing houses in Italy and the world-famous opera House, La Scala. The city's varied industries include:—

textiles, aircraft, electrical goods, food processing, railway equipment, steel, machine tools, chemicals, vehicles.

TURIN (1 120 000) is also a focus of communications. (15) Explain why, referring to the map opposite and to your atlas. In former times Turin was chiefly important as the capital of Savoy, a small kingdom which wielded considerable influence on account of its position astride the Alpine passes between France and Italy. Since 1899, however, the fate of Turin has been inextricably tied to that of the great firm of FIAT, which now employs over 80 000 workers. In addition to vehicles (p. 154), FIAT also produces a great variety of light and heavy engineering products, e.g. railway equipment, aircraft and machine tools. Turin also produces textiles, vermouth and leather goods.

GENOA (850 000) is an ancient port, built in terraces round a deep-water bay. It is the main port of Northern Italy and also handles much Swiss trade.

Major industries in Genoa include: *shipbuilding, engineering, petroleum refining* and *steel-making* (20% of Italy's total output of steel comes from an enormous steelworks in the industrial suburb of Cornigliano). Genoa's future as a major port is uncertain. Difficulties include (i) congestion and delays in the docks, (ii) declining liner traffic since the blockage of the Suez Canal and (iii) ever-growing competition from Rotterdam (see *page 116*) and to a lesser extent from Antwerp and Hamburg. In a bid for economic survival Genoa has built the first container docks in the Mediterranean.

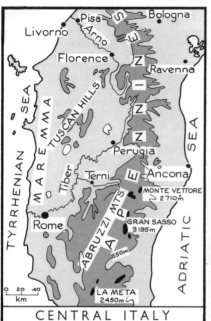

Gully erosion in the Apennines.

CENTRAL ITALY lies between the prosperous, industrial 'North' and poverty-stricken, agricultural 'South'. Within the region living standards vary greatly. More than half the inhabitants are peasant farmers, but there are some large towns, e.g. Florence, with flourishing industries. Many of the country folk, especially those living on the eastern side of the peninsula, live a hard, frugal existence. Farmers of the fertile basins and plains farther west, however, are much better off. Generally speaking the people of Central Italy are fairly prosperous, mainly because the density of population is lower than elsewhere.

There are marked contrasts in the appearance of the land and in the uses to which it can be put on opposite sides of the Apennines 'backbone'. (*(16) Explain the use of the term 'backbone' in this context. Name, and state the altitude of, the highest parts of these mountains.*) The eastern foothills rise sharply from the Adriatic Sea, leaving very little coast plain. The rocks are rather soft sandstones, clays and limestones, and the photograph shows how deeply they have been eroded by large numbers of turbulent streams. Flat, fertile land is obviously scarce, and the valleys are avoided because they are choked with sand and liable to sudden flooding. In their efforts to produce arable land, generations of peasant farmers have cut down the oak forests which once clothed these ridges. As a result the rain-water from torrential winter storms rushes downhill, washing away the soil and adding to the

farmers' difficulties. Tree crops, e.g. olives and vines, have been planted to help remedy the damage. These, together with wheat grown on small terraces cut into the hillsides, are the main products of this part of Italy.

Between the Apennines and the west coast there are a number of hills and undulating plateaus. (*17*) What are the names (*a*) of this hilly region; (*b*) of the narrow plain along the coast; (*c*) of the two main rivers entering the Tyrrhenian Sea? These rivers drain several small basins containing fertile alluvial soils. The fertile lower ground is intensively cultivated; wheat, vines, maize, tobacco and olives are the main crops—a typical Mediterranean 'polyculture'. Enough fodder is available for cattle to be kept in fair numbers—an unusual feature in Peninsular Italy.

The hills, still partly forested, are made of a great variety of rocks including sandstones, limestones and lava. The presence of the latter, and of steam vents, crater lakes and hot springs, remind us that there were active volcanoes in Tuscany and Umbria until quite recent geological times. The whole region is still frequently shaken by earth tremors.

Details of **geothermal power** output in this part of Italy are given on page 153. At Lardarello there is also an important chemical works which extracts impurities from geothermal steam and hot water. Its main products are carbonic acid gas, borax and sulphur.

CENTRAL ITALY: Relief, Major Towns, Routes, Apennine Passes and Mineral Deposits

The Northern Apennines rise abruptly from the west coast, reaching over 2350m in a series of flat-topped ridges and domes. They present a formidable barrier to north-south traffic, but are crossed by several important road passes and pierced by railway tunnels.

(*18*) Name the passes used by the following routes:—

(*a*) Bologna–Pistoia; (*b*) La Spezia–Parma; (*c*) Bologna–Florence; (*d*) Genoa–Milan; (*e*) La Spezia–Reggio.

(*19*) Which river valleys are followed by traffic between Florence and Rome? This route follows the path of a famous Roman road, Via Flaminia.

Livorno (*Leghorn*) (170 000) was developed as a port during the 16th century. It soon took over the trade of the nearby port of Pisa, which was badly affected by silting. Today Livorno has chemical, engineering, oil-refining and shipbuilding industries.

Elba contains Italy's largest deposits of iron ore, and produces about 1 m. tonnes annually. The ore is sent for smelting to Genoa, Piombino and Follonica.

The West Coast consists of a series of shallow, sandy bays, fringed with sand bars and swampy mud-flats. Until recently these were largely uninhabited and infested with malarial mosquitoes. During the past forty years large-scale reclamation has been carried out, and the reclaimed land has been divided into peasant holdings. The formerly desolate Pontine Marshes, e.g., now support a farming population of over 20 000. Crops include wheat, vines, fruit, hemp and cotton.

Florence (460 000) is an important route focus, lying on the vital road and rail links between Rome and the Po Basin. For centuries the city has been an outstanding centre of art and culture and it contains many famous buildings and sculptures. The main source of income for the town's population is tourism, but there are also chemical, engineering, railway and electrical industries.

The Hills of Tuscany contain fair quantities of a variety of minerals, many of which are of volcanic origin. Products from Tuscan mines and quarries include: sulphur, gypsum, salt, lignite, tin, mercury and borax.

Narni and *Terni*, at one-time small market towns, have recently acquired important cement, chemical, steel, engineering and aluminium-refining industries. Power is obtained from large hydro-electric stations on the Nera River. ((*20*) Of which river is the Nera a tributary?)

The Abruzzi contains the highest and most formidable of the Apennine ranges, which in places exceed 3000 m. The main rock is limestone, and there are large areas of barren, glaciated karst plateaus. The high ground is snow-covered for up to five months, but is used in summer by thousands of transhumant sheep. Transhumance has been practised here since Roman times, and centuries of browsing and trampling have caused much deforestation and soil erosion.

MILAN

PO BASIN

La Bochetta
Genoa

La Cisa

Parma

Cerreto

Reggio

La Spezia

MARBLE SULPHUR Bologna

Pisa Reno

Livorno Arno Pistoia La Futa

Lardarello Florence

Piombino

TUSCAN HILLS
SULPHUR, SALT
GYPSUM, ALABASTER, COPPER
BORAX etc.

ELBA
Follonica

LIGNITE

PYRITES

TIN
MERCURY
MTE. Amiata

Tiber

Narni
Terni

Nera Gran Sasso

ROME ABRUZZI

PONTINE MARSHES

0   50
km

NAPLES

The Pope blessing vast crowds in St. Peter's Square, Rome.

Details of other minerals produced in Central Italy are given on the map opposite. The total output from mines and quarries is comparatively small, but they add to the wealth of the region and provide some alternative employment to peasant farming. So do factories in such large towns as Florence, Terni, Leghorn and Rome. The result is a steady drift of younger people from the countryside, especially from the remoter parts of the Central Apennines where life is often hard and unrewarding. For those seeking the 'bright lights' and a chance to get on in life the chief magnet is Rome—the great city which has dominated the life of Central Italy for over 2000 years.

ROME (2 495 000) grew up where five main routes converge to cross the Tiber at its best bridging point (see map). Throughout the period of the Roman Empire the city collected grain and wealth as tribute from all the conquered territories. This made Rome independent of supplies from the surrounding countryside, and thus allowed its population to grow rapidly. The Empire fell in the 5th century A.D., but Rome, as the heart of Christendom, continued to draw tithes (taxes) through the Church. Pilgrims and tourists still flock to Rome, and the money they spend is one of the main sources of income for the city's inhabitants.

When modern Italy finally became unified, Rome's enormous prestige and central position in the country made it the obvious choice for capital. 'All roads lead to Rome'—and in modern times all Italian railways do the same. The city also has the busiest airport in the Mediterranean area, being a focus for flights from N.W. Europe, East Africa and the Near and Far East. In recent years Rome has acquired a variety of industries, of which the most important are food-processing and railway engineering. Craft industries also produce a great number of tourist souvenirs. Rome is the main centre of the Italian film industry, and a leading European centre of fashion and luxury clothing production. Government offices, international organisations (e.g. F.A.O.) and commercial agencies in Rome provide work for many thousands of administrators and clerks.

THE SOUTH

Geographically the South consists of the five administrative regions of Southern Italy; but it is often taken to include Sicily and Sardinia, which have shared its economic backwardness and grinding poverty for so long.

"While Italy's southern peasant thinks of Rome and Milan as misty Utopias fully as remote as America, the man-in-the-street Northerner thinks of all Italy south of Rome as a barbarous land in which civilised standards are low and the inhabitants are unkempt, primitive people. Both images are exaggerated, but the second less than the first. There are villages in Sicily and Campania where people live with their donkeys in caves. In the city of Naples, scattered underground, colonies of humans, living like moles, cough up their tubercular lungs. In superstition-ridden Apulia, the heel of the Italian boot, there are people who dance the wild *tarantella* to cure themselves of the tarantula spider's bite. In desolate Calabria some villages have no roads and are approached by way of the stony beds of streams. In Sardinia, bandits gather high on a bleak hill for 10 days each May to pray to their patron, St. Francis, asking his protection for their work. In Benevento children are auctioned for labour on farms. At Palma infant mortality is 10 times that in the North, and if you ask a mother how large her family is she may reply, 'We are twenty—ten living and ten dead.'

"Yet these are the lands the Greeks settled 2,700 years ago and made a centre of culture. . . . How does it come about that this former centre of civilisation should be in such despair today? The answer is a tangle of history and geography. But one cause rises above all others: deforestation." (Kuhly, *op. cit.*)

When the Greek colonists arrived in Southern Italy they found the whole region densely forested with oak, laurel, ilex and myrtle. Today hardly a tree remains—the result of nearly 3000 years of indiscriminate felling for fuel, shipbuilding and charcoal. Close-cropping of grass and young saplings by generations of sheep and

Acerenza, a typical hill-top village in Basilicata. Such sites were chosen for defence and to avoid malarial-infested lowlands. Note the vineyards and evidence of gully erosion. The entrances on the lower hillslopes lead into cave dwellings.

goats completed the devastation. Many of the worst afflictions of the South—soil erosion, sudden 'flash' floods and rock avalanches —stem directly from deforestation. The floods in turn produce swamps and stagnant pools on the lower ground, once breeding spots for the *anopheles* mosquito which spread the South's former scourge, malaria. To this add occasional devastation by volcanic eruption and earthquake (over 100 000 people died in the great Messina earthquake of 1908), and centuries of banditry, warfare and pillage, and it is not difficult to understand why the people of Southern Italy are amongst the most backward in Europe. For example, recent investigations reveal that 40% of the 20 000 inhabitants of Palmi di Montechiaro in Sicily can neither read nor write, and 3000 of them live five to a room. Many of the rooms are windowless and shared with goats or mules. Four out of every five houses have neither water nor sanitation. Virtually all the children suffer from worms and many of them have the dreaded eye disease, trachoma. The daily diet consists of bread, spaghetti, oil and a few vegetables. Meat is eaten as a rare treat only two or three times a year.

Farming provides a meagre livelihood for the vast majority of the South's 21 million inhabitants, and defects in agriculture are the root cause of the widespread poverty and distress. In many parts of the South there are enormous private estates, **latifondi**, where the system of land ownership and farming methods are still much the same as they were in the Middle Ages. On these estates about a third of the land is allocated to **share-croppers**, each of whom works a tiny plot to grow subsistence crops for his family in his spare time. The rest of the estate is divided into large fields and worked as one unit by the share-croppers. In return for their labour the share-croppers are provided with seed, fertilisers, tools and draught animals by the owner, and receive a small share of the proceeds when the owner's crops are sold. Wheat is usually the main crop, grown in rotation with beans. Day-labourers (*braccianti*) are also employed on the estates. Most share-croppers and *braccianti* live in large, slum-ridden, hill-top villages from which they walk as much as 13 kilometres each day to their place of work. Grave defects arise in this system because:

(i) owners usually do not live on their estates; they appoint agents to run the *latifondi* and take little real interest in farming.

(ii) share-croppers' plots are too small to assure a decent living, and payments for work on the owners' fields are very small.

(iii) *braccianti* are often unemployed because they are used only for seasonal work such as sowing and harvesting.

(iv) there are far too many peasants trying to get a living from the small amount of fertile land available.

Near large towns, and in places where there is water for irrigation, e.g. Northern Sicily and Western Calabria, specialised cash crops such as citrus fruit, vegetables and flowers are grown on irrigated small-holdings owned by peasant proprietors. They are so small, and families are so large, that the owners often find it hard to make ends meet.

Until the 1920's the overpopulation problem of the South was eased by the annual emigration of scores of thousands of peasants, mostly to North or South America. Then the U.S.A. imposed strict immigration laws, and this 'safety valve' was largely shut. In the 1930's there was an outlet to the Italian colonies in Africa, but Italy lost these territories during the Second World War. Details of how the Italian Government is now trying to solve the South's problems are given on page 171, but first we must learn more of the geography of the region.

Wash day in the congested slums of Palermo.

SARDINIA

COSTA SMERALDA

R. Coghinas

LA NURRA

Olbia

Sassari

FERTILIA

Oristano

R. Tirso

ARBOREA

R. Flumendosa

CAMPIDANO

Iglesias

Carbonia

Cagliari

SARDINIA consists largely of hard crystalline rocks with very thin and infertile soils. The island's mountainous interior has a wild, rugged and desolate appearance and is mostly uninhabited.

Scattered flocks of goats and sheep graze on the poor upland pastures, but nearly all of Sardinia's 1 500 000 inhabitants live in the more fertile river basins and narrow coast plains. For centuries pastoral farming was virtually the only source of livelihood, but recently attempts have been made to increase the area under cultivation. (*21*) Draw a *large* outline map of Sardinia. (*22*) Label the interior 'Poor Pasture Land', and add other labels (using arrows where necessary) to show that:—

1. Olives are grown in large numbers around Sassari.
2. Grain-growing is important in the Fertilia Lowlands.
3. The lowlands of Arborea is the main cultivated area. Crops include wheat, maize, rice, tobacco and greenhouse products.
4. Vines and wheat are grown in the Campidano.
5. H.E.P. and irrigation works have recently been constructed in the valleys of the Rivers Tirso, Coghinas and Flumendosa.
6. Lead, zinc and iron ores are mined in La Nurra and at Iglesias. Coal is mined at Carbonia and used to generate thermal-electricity.
7. Oristano and Cagliari have food-processing industries. Olbia is the main ferry-port, with regular sailings to Civitavecchia.

LAND > 215m

0   40   80
km

Trapani

Palermo

Cefalu

Messina

Marsala

ETNA 3580m

Catania

Porto Empedocle

Augusta

Gela

GAS & PETROLEUM

Ragusa

SICILY

SICILY consists mainly of barren hills and low plateaus: limestone karst covers 27% of the island.

—has only narrow coastal plains, except round Catania.

—suffers severe summer drought and loses much winter rain-water by evaporation and rapid run-off.

—is subject to severe earthquakes and frequent landslips.

—occupies a strategic position at the crossing of land and sea routes ((*23*) explain) and so has a long history of invasions, warfare and pillage.

The vast majority of the island's 5 million inhabitants are supported by agriculture, hard wheat being the staple subsistence crop. Yields are amongst the lowest in Italy, and drought causes frequent crop failures. The most favourable agricultural districts are (i) the narrow coast plain between Palermo and Cefalu and (ii) the Plain of Catania and lower slopes of Mt. Etna. The latter is an active volcano and the ash soils which cover its lower slopes are remarkably fertile. In both districts tree-crops (olives, almonds, vines), citrus fruit (especially lemons), vegetables and flowers are grown for export. Water for irrigation is obtained from springs flowing from the flanks of the limestone hills, and from streams and wells. The droughty Plain of Trapani is well known for its strong Marsala wines.

Sicily has few industrial raw materials except sulphur, but urgent efforts are being made to build up industries to help relieve the unemployment and poverty which afflict so many of the island's population. The main hope lies in developing the recently discovered petroleum and natural-gas fields (*see map*). Oil-wells at Gela and Ragusa already produce over 2 million tonnes per annum. In addition many new enterprises have been established since 1948 with government assistance. They include oil refining and petro-chemicals, steel-making and precision engineering, textiles (including fibre made from Etna lava) and fertiliser and chemical works. The main industrial centres are Palermo (*food-processing, shipbuilding and ship repairing, textiles*), Porto Empedocle (*chemical fertilisers*) and Augusta (*oil refining*). Messina and Catania are both busy ports where agricultural produce is assembled for export.

Using the information in the previous two paragraphs (*24*) make a *large*, labelled sketch-map to show the economic activities of Sicily.

View across Naples harbour to Mt. Vesuvius. The ridge on the left is Mt. Somma.

PENINSULAR SOUTHERN ITALY is mostly bleak and barren, with limestones massifs rising abruptly from the coast to over 2000 m. Large areas are treeless, waterless and useless except as poor sheep pasture. A few places, however, have sufficient water from springs, rivers and reservoirs to allow irrigated polyculture. The main cultivated areas are (i) the coastal plain and lower hill slopes around Naples and Salerno, including the highly fertile volcanic soils on Mt. Vesuvius and (ii) the coast plain around Bari, where the soils are rich in natural phosphates. (25) From which river is water brought to Apulia by aqueducts and tunnels through the Apennines? Other patches of intensively cultivated land are indicated on the map. A great variety of crops are raised in these arable 'oases', including maize, wheat, nuts, mulberries, vines, vegetables, olives, figs and citrus fruit.

The population of peasant farmers and casual farm labourers is so dense, however, that average incomes are pitifully low and there is much unemployment, in spite of the prosperous appearance of the orchards and fields. As in the northern part of the Peninsula (see page 163) there is a steady drift of dissatisfied young people from the countryside to the few large towns.

The Straits of Messina. Ferry boats plying between Messina and Reggio carry most of the fruit and vegetables despatched by Sicilian farmers to markets in peninsular Italy. Proposals exist for the construction of a suspension bridge to by-pass this bottleneck.

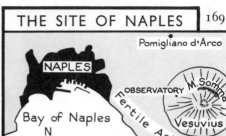

THE SITE OF NAPLES

NAPLES

Pomigliano d'Arco

OBSERVATORY

M. Somma

Vesuvius

Fertile Ash Soils

Bay of Naples

N

0   5   10
km

Torre Annunziata

every year to seek work in the North or in Bari—the " rising star of the South ". The Naples Development Society proposes the building of a new town inland at Pomigliano d'Arco where the new Alfa-Sud vehicle factory already provides a livelihood for more than 5000 families. (Photo p. 173)

Naples (1,192,000) is the only major port and large-scale industrial city in southern Italy. It occupies a sheltered, south-facing site, with magnificent views of the coast and the towering Mt. Vesuvius. There are regular sailings of passenger liners from Naples to the Americas and to East Africa, the Far East and Australia. As a commercial port Naples ranks a close second after Genoa, importing large quantities of industrial raw materials, notably:—

| petroleum | cotton | phosphates |
| metal ores | jute | coal |

The main industries of Naples are oil-refining, steel-making and engineering, shipbuilding, textiles and food-processing (e.g. fruit- and vegetable-canning and olive-oil extraction). There is a string of industrial suburbs reaching southwards around Naples Bay to Castellammare, and textiles are manufactured in Salerno. Naples is also an Italian and N.A.T.O. naval base and a world-famous tourist centre.

In spite of its characteristic noise and bustle Naples reflects the backwardness of the South in its teeming slums and in the desperate poverty of many of its citizens. Widespread unemployment is aggravated by an annual population increase of 40 000, and some 10 000 men leave Naples

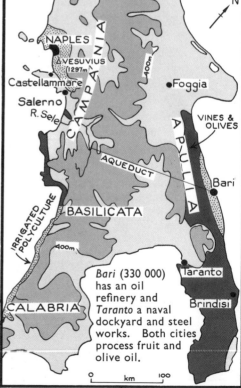

PENINSULAR SOUTHERN ITALY

N

NAPLES

VESUVIUS
(1297m)

Castellammare

Salerno

R. Sele

CAMPANIA

Foggia

1000m

VINES &
OLIVES

APULIA

AQUEDUCT

Bari

IRRIGATED POLYCULTURE

BASILICATA

400m

CALABRIA

Taranto

Brindisi

Bari (330 000) has an oil refinery and *Taranto* a naval dockyard and steel works. Both cities process fruit and olive oil.

0   km   100

This photograph shows the country west of Catania where large irrigation schemes are being implemented. In the foreground is a water storage tank and new house. The reclaimed land around the house has been planted with vines. In the distance, at the foot of the mountains, is Nova Catena: this small town hqs been built since 1956 in conjunction with the land reclamation scheme.

Re-development in the South is the most pressing problem facing Italy today.   Just after the Second World War conditions amongst the *braccianti* became so bad that restless bands of them marched on estates and threatened to take possession of idle land by force. Parliament rushed through a land reform bill, breaking up inefficient *latifondi* and distributing the land to peasant farmers. Some estate-owners also relinquished their land voluntarily, and so far over 100 000 families have been re-settled on small farms throughout the South.   To help the new owners (most of whom are unskilled and illiterate) the Government arranges instruction in such modern farming techniques as deep ploughing, drainage and tractor-driving.   Special co-operatives have also been set up for harvesting, marketing, wine-pressing, olive-oil milling, tomato-canning and so on.

In 1950 the Government established the 'Southern Italy Development Fund', to improve agriculture and foster industrial growth.

Between 1950 and 1970 £2 100 000 000 was allotted to the Fund.   Over 70% of this has been spent on agriculture and the remainder mostly on aqueducts, sewage disposal, roads, railways and the tourist trade.   The total list of achievements to date is impressive (*see table, right*), but much remains to be done if the gap in living conditions and social amenities between South and North is to be closed. About 15 million people in the South, i.e. three in every four, are still directly dependent on agriculture. Many farming communities contain far more people than can profitably be employed on the land, and it is estimated that at least 8 million new jobs must be created during the next 10 years. Only in this way will it be possible to eliminate the bitter poverty of the remoter country districts and to stem the flood of c. 300 000 disillusioned migrants who at present leave Southern Italy every year.

Since the renewal of the *Cassa* in 1965 much greater emphasis has been placed on the establishment of industry in Southern Italy, supported by an **infrastructure** of such varied services as roads, railways, ports, banks, financial institutions, schools, technical colleges, machine-tool factories and repair workshops. The aim is to establish 'poles of expansion', where growing industries and service trades will absorb an ever-widening circle of employees. Special attention is being paid to creating industrial development zones at Bari, Brindisi and Taranto. At Taranto, for example, the *Cassa* has established Italy's largest single iron and steel plant—one of the biggest in Europe, employing 5000 men and with an annual output of 5 million tonnes. Special jetties have been built to handle imported iron ore and coking coal, and a new port area, shipyards, industrial estates are all being planned. The growing range of engineering works already in Taranto which use the locally produced steels is indicated on the right. Bearing in mind that Taranto also has an oil refinery, cement

Main Achievements of the Southern Italy Development Fund (*Cassa per il Mezzogiorno*)
Construction of:— 30 000 km of roads 13 600 km of canals 12 000 km of aqueducts 2 400 km of dykes
Digging of over 20 000 wells; water made available to over 8 million people.
Agricultural development schemes including the redistribution of 700 000 hectares of land and construction of 60 000 farmhouses and 200 hamlets. New agricultural institutions run courses on tree culture, animal care and breeding, soil and farm management.
New developments sponsored in forestry, fishing, archaeology and tourism throughout the South.
Construction of the *Autostrada del Sole*, a fine modern motorway cutting through the mountains by spectacular viaducts and tunnels. Reggio Calabria now linked to Rome and to the markets of the North. An offshoot links Bari to Naples and a new autostrada from Bari to Bologna along the Adriatic coast will be ready by 1972.
Establishment of heavy industry at Taranto, Bari, Brindisi, Gela and Cagliari. (*For details see text*).

Some Typical Products of Engineering in Taranto
pumps agricultural machinery mechanical excavators cranes machine tools structural steelwork elevators domestic appliances

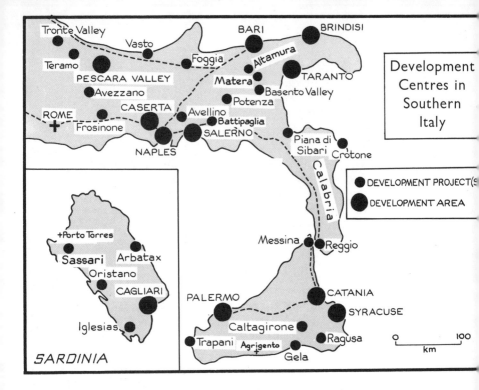

Development Centres in Southern Italy

- ● DEVELOPMENT PROJECT(S)
- ⬤ DEVELOPMENT AREA

SARDINIA

works and a shipyard, (26) what do you think is meant when Italians talk of Taranto as being a '*catalyst*' for the entire South?

All Italian state-controlled enterprises are now required by law to locate 60% of all new investment in the South. Such enterprises include *Italsider* (iron and steel), *E.N.I.* (oil and gas) and *Breda* (mechanical engineering). Private firms are also moving into the South in greater numbers, the *Cassa* having attracted more than 500 businesses into the Rome-Naples area alone. The table shows the principal industrial developments of the Italian South: (27) put this information on a labelled sketch-map and give your completed map an appropriate title. So far the *Cassa* has provided jobs for about 500 000 people, but even so parts of the South still lag sadly behind in economic progress. In Calabria,

---

Main Industrial Centres in Southern Italy

Oil refining and petro-chemicals:

BRINDISI	TARANTO
CAGLIARI	SYRACUSA
PORTO TORRES	FERRANDINA
	(Bassento Valley)
GELA	BARI

Metallurgy: PORTOVESME

Engineering: TARANTO BARI
BATTIPAGLIA
BRINDISI

Iron & Steel: TARANTO NAPLES

Vehicles: NAPLES (*Alfa-Sud*)
PALERMO (*Fiat*)

Shipbuilding: MESSINA
PALERMO
NAPLES

Food Processing: BARI
CATANIA
PALERMO
NAPLES
ALTAMURA
BATTIPAGLIA

172

The Alfa-Sud car factory at Pomigliano d'Arco, at the foot of Mt. Vesuvius. This works is one of the new industrial 'growth centres' which it is hoped will bring economic prosperity to Southern Italy.

for example, there are still only four factories with more than 500 employees, and the average wage in the Agrigento region of Sicily is a mere £89 per annum. Many argue that much more should be done to stimulate industries such as fruit and vegetable canning, wine making and distilling, vegetable-oil extraction and fishing, i.e. industries which are rooted in the traditional way of life of the *Mezzogiorno.** Enterprises such as the new cattle breeding station in Catania (where a special breed of dairy cow has been evolved to withstand the heat and drought of the Mediterranean climate) meet with approval, but new industries such as petrochemicals and engineering are resented because it is feared that they will shatter the unique Southern cultural heritage. (*28*) Do you agree with this point of view?

* *Mezzogiorno* ('Mid-Day') is the name by which the South is known in Italy. (*29*) Suggest a probable origin for this term.

The main hopes of raising living standards in Italy depend on expanding industry and foreign trade. The table below reveals two facts about the nature of Italian trade:—(i) *Italy has to rely on imports for certain vital commodities, e.g. fuels and basic foodstuffs, but many of her exports are of semi-luxury goods such as citrus fruit, wines and motor scooters.* This means that when times are hard, foreign customers can cut down on imports of Italian goods more easily than Italians can restrict their overseas spending. (ii) *The money spent by Italians on imports is much more than is earned by selling Italian goods overseas.* This 'trade gap' is partly bridged by the earnings of Italian ships and aircraft carrying goods and passengers for foreign countries, and by money sent home to relatives by Italians working or living overseas. Both these sources of foreign revenue have declined in recent years, but another source, tourism, has vastly increased.

Tourism is now one of the pillars of the Italian economy, employing as much labour as the largest industrial undertakings and bringing in about £400 million in foreign currency every year. The number of tourists flooding into Italy annually to spend this money is some 20 million—i.e. considerably more than the total population of Greater London. Their spending is particularly valuable because it is spread throughout the country, including the impoverished South. The money also goes to a great number and variety of people, e.g. hotel and cafe owners, shop-keepers and stall-holders, guides and entertainers. The Italian Government has indicated the importance it attaches to tourism by setting up a Ministry of Tourism and Entertainment. Many hotels have been built with government aid, and efforts are being made to popularize such backward regions as Apulia, Calabria, Sicily and Sardinia. Tourist spending has brought about a revival in traditional craft industries, e.g. the making of gloves in Naples, silk in Milan, embroidery and lace in Venice, Sicily and Tuscany, and straw and wicker-work goods throughout the country.

THE FOREIGN TRADE OF ITALY

Main Imports	Main Exports
Foodstuffs (e.g. food grains, animal products and tropical fruit)	Foodstuffs (e.g. lemons, oranges, olive oil, cheese, canned fruit and vegetables)
Forest Products Coal Metal Ores	Wines Textiles Clothing
Petroleum Metal Goods and Machinery	Vehicles Metal Goods and Machinery
	Chemicals Rubber
Total Value of imports £5090 m.	Total Value of Exports £3150 m.
The main countries sending goods to Italy are:—U.S.A., W. Germany, France, U.K., Kuwait, Netherlands, Belgium and Argentina	The main countries to which Italian goods are sent are:— W. Germany, France, U.S.A., Switzerland, U.K., Netherlands and Belgium

The most famous of all Italian tourist centres, VENICE.

VENICE was founded in A.D. 468 by a group of merchants who sought safety from the barbarian invaders of Italy. (*(31) What advantages did the site of Venice offer in this respect?*) In the Middle Ages Venice became the main port of exchange between the Western world and the East, and was for centuries the most powerful city in Europe. Goods landed at Venice were distributed throughout the Continent. (*32) Suggest why decline set in for Venice following the discovery of America and the sea route to India via the Cape of Good Hope.* More recently the increase in the size of ships has hindered the trade of Venice (*(33) why?*), but she is still the third port of Italy. (*(34) Name the first two.*) Tourists are drawn to Venice ('The Queen of the Adriatic') to gaze at her beautiful buildings and priceless art treasures, to explore the canals and to enjoy the beaches and water sports of the Lido. Apart from an arsenal, the only important industries in Venice are tourism and the related handicraft trades, e.g. glass, silverware and lace. The map shows the several important industries that have grown up at the new town and port of nearby Mestre.

THE SITE OF VENICE

Metallurgy
Chemicals
Shipyards
Oil Refining

Mestre

DOCKS

VENICE · Lido

LAGOONS

SAND-DUNES

SAND-DUNES

Adriatic Sea

LAGOONS

Chioggia

R. Adige

R. Po

PO-DELTA

0 ___ 15
km

Other well-known resorts are located between San Remo and Viareggio on the 'Italian Riviera'—the eastward extension of the 'French Riviera' described on page 74. (*30*) Where in Italy would you go if you wished (*a*) to see (i) the Vatican, (ii) La Scala, (iii) the Doge's Palace, (iv) the Palio, (v) Pompeii, (vi) Michelangelo's painting 'The Last Supper', (vii) St. Peter's; or (*b*) to (i) peer into the crater of an active volcano, (ii) climb Alpine peaks, (iii) trek over wild granite moorlands?

CHAPTER 8

Southern Europe: (ii) Spain, Portugal and Greece

(1) SPAIN

"At 6.30 in the morning, when a layer of mist still hung over the palm trees of the flat valley, I sat sleepily in a little four-wheeled railway carriage which climbed up away from the city of Murcia along the edge of some hills, pulled by an old steam engine. The carriage was filled with country-folk chattering away to each other. One voice could be heard continuously above the others: 'You know, they say . . . that it is farther from here to the moon . . . than from here to Madrid.' A long silence followed this revelation. Then someone else said sceptically: 'I don't believe you.'"*

This conversation is not quite so extraordinary as might at first appear. These maps show that the people of Iberia† live in clusters—on patches of coastal plain and in river basins—isolated by great stretches of barren, sparsely inhabited mountains and plateaus. Many of them are ill-educated peasant farmers, who rarely travel far from their villages and farms. Thus they have little knowledge of, or interest in, the world beyond their particular valley. Local patriotisms are very strong, and Spain has been plagued for centuries by internal jealousies and sometimes warfare between different regions. Attempts by politicians in Madrid to impose their authority on remote cities such as Barcelona and Bilbao are bitterly resented. The disunity of Spain was illustrated by the terrible Civil War of 1936-9 in which the country was devastated and half a million Spaniards lost their lives. In particular the desire of the Basque people of north-east Spain to have a greater degree of independence is a long-standing and bitter internal feud. Unity is still hindered by the rugged terrain and by notoriously poor communications.

* Alexander Rainey, the *Geographical Magazine*.

† Spain and Portugal are known collectively as Iberia

(1) Suggest, with reasons, which set of figures belongs to each of:— Corunna? Madrid? Alicante? (Refer, if necessary, to pp. 14–18). (2) Name the three types of climate represented, and compare with the map on page 18. Although the seasonal distribution and total amount of rainfall varies considerably from place to place, most of the peninsula is very arid (map overleaf).

		J	F	M	A	M	J	J	A	S	O	N	D	Year
A	°C	4	6	9	12	16	21	24	23	19	14	8	6	—
	mm	28	43	43	43	38	30	10	8	30	48	55	40	416
B	°C	9	10	11	12	14	16	18	18	17	14	12	11	—
	mm	80	78	80	63	55	35	23	30	55	88	105	110	802
C	°C	12	12	14	16	19	22	25	26	23	19	16	13	—
	mm	30	23	18	33	18	3	3	10	20	18	33	38	247

IBERIA: Relief, Rivers and Main Towns

(1) Which mountains, plateaus and main rivers would you cross on a journey to Madrid from (a) Malaga, (b) Barcelona, (c) Bilbao, (d) Gijon and (e) Cadiz?

(2) Which mountains virtually seal off the Iberian Peninsula from the mainland of Europe?

(3) Of which fold mountain chain are the Balearic Islands probably detached remnants? (*Map p. 8.*)

(4) In which direction is the Meseta tilted? How do you know?

(5) Which river draining to the Atlantic does *not* flow through Portugal?

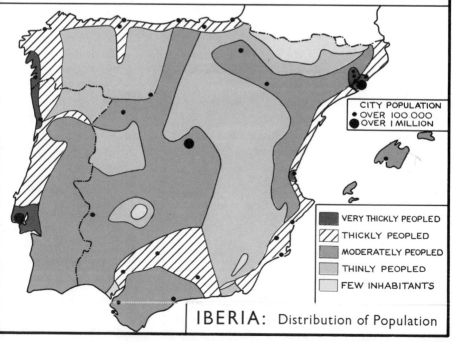

CITY POPULATION
● OVER 100 000
● OVER 1 MILLION

VERY THICKLY PEOPLED
THICKLY PEOPLED
MODERATELY PEOPLED
THINLY PEOPLED
FEW INHABITANTS

IBERIA: Distribution of Population

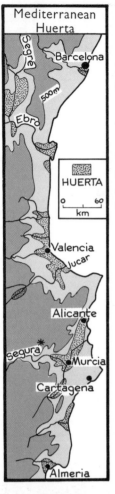

Mediterranean Huerta

Where possible, irrigation is practised throughout Iberia, for even in the comparatively rainy north and north-west the evaporation rate in summer is very high.

THE MEDITERRANEAN COAST of Spain contains the most productive irrigated lands in the Peninsula. Life-giving water from the small but vitally important rivers shown on the left is run through an intricate system of pipes and channels to intensively cultivated *huertas* (gardens). The main *huerta* districts are shown on the map, and the following description is typical:—

"Hemmed in on either side by thousands of square miles of semi-desert, scrub and badlands runs a valley, seventeen miles long and about five wide, rich almost to the point of exaggeration. Steep, sharp mountains on one side seem to dive headlong into this valley to re-emerge the other side. This is the *Huerta* (vegetable- and fruit-growing land) of the Spanish province of Murcia, in the middle of which stands the provincial capital, Murcia, shrouded with the sweet scent of orange blossom. Every inch of it is minutely cultivated with extreme care and hard work. There is an abundance of fruit trees, of almonds, figs, dates, oranges, apples, pears, pomegranates, quinces, grapefruit and, above all, lemons (largest production in Spain), not to mention other crops: corn, maize, cayenne pepper, olives and even rice, and many different vegetables; in fact, almost anything you could think of that is eaten by Europeans. Production is not limited to foodstuffs. 45 000 acres are used for cotton-growing. Four-fifths of the Spanish silk production comes from here, 2500 families have the special mulberry trees . . . were it not for the River Segura* this area would be a desert . . . the sun, blazing down from a

* This river will soon gain extra water brought from the Tagus via 240 km of canals, siphons and tunnels.

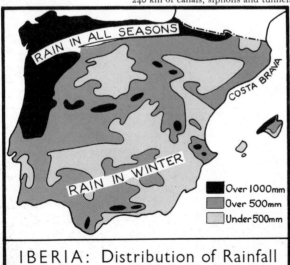

IBERIA: Distribution of Rainfall

Over 1000mm
Over 500mm
Under 500mm

BARE MOUNTAIN
TERRACES
Tree crops

Coastal Plain
IRRIGATED HUERTAS

Reclaimed Lagoons
↓
RICE

Sandspits & Dunes
↓
Medit.

perpetually blue sky and reflected off the rocks of the badlands, knocks the temperature up to 114°F. now and again. Sometimes there is a 'wind of fire' from the Sahara, which is so hot that one day in the shade my glass was hot a moment after I had drunk down an iced Coca-Cola." (A. Rainey, op. cit.)

The irrigated *huertas* lie mostly along river valleys and on narrow patches of alluvial coast plain, but the diagram above shows that the lower mountain slopes are also terraced and cultivated. Most of the peasants who work in the *huertas* live in large, widely spaced villages, so that the land looks strangely unoccupied despite its dense rural population. Except for the orange groves, which are usually owned by larger proprietors, *huerta* plots are very small, and there is the same desperate pressure of people on limited supplies of cultivable land as in Southern Italy.

Much of the agricultural produce of this part of Spain is exported, especially oranges, tomatoes, onions, artichokes and other kitchen vegetables. Cereals, tobacco and fodder crops are grown for local consumption. Large quantities of foodstuffs are also supplied to the cities shown in the map opposite, both for consumption and for food-processing industries. By far the largest of these cities is Barcelona.

The Murcia huerta.

BARCELONA'S industrial development was favoured by (*a*) its coastal location (*fuel and most raw materials for Spanish industry had to be imported*) and later by (*b*) ample supplies of power from running water, a scarce commodity in Spain. The area now depends increasingly on h.e.p., much of it generated in Lerida Province. The main industries of Barcelona are textiles, shipbuilding, engineering, electrical equipment, diesel engines, vehicles, etc.) and chemicals. Many firms in Barcelona are associated with foreign concerns; e.g. SEAT vehicles use parts manufactured by FIAT of Turin. Liner routes operating from Barcelona are indicated on the map, together with details of other industrial towns served by the port.

The Barcelona Region

Barcelona (1 640 000) is the most important commercial and industrial centre in Spain. The city's fame dates back to medieval times, when it was already a great trading port. In the late 19th century, when Spain first took hesitant steps towards industrialisation, Barcelona's merchants used their business enterprise and wealth to foster the growth of industry in the city.

Also important is *Valencia* (510 000), the third city of Spain, through which great quantities of locally produced fruit, wine, raisins, olive oil and iron ore (from nearby Teruel) are exported. Valencia also has food-processing, textile, leather, engineering, chemical and shipbuilding industries. *Murcia*, *Alicante* and *Malaga* are similar regional centres of *huerta* 'oases'.

Tossa. Costa Brava.

The whole Mediterranean coastline of Spain is enjoying a great boom in tourism. Spain has always been favoured by British and other North European holiday-makers, but until recently it was relatively inaccessible. Cheap air transport and the ever-increasing number of privately-owned cars now allow tens of thousands of foreign visitors to book a Spanish holiday and be certain of finding 'a place in the sun'. (See photo.)

The best-known Spanish Mediterranean resorts lie on the Costa Brava (map p. 178) & the Costa del Sol (map p. 182). This coast, together with the colourful interior cities of Cordoba, Seville and Granada, is undergoing very rapid development which may soon make it Europe's most important region for tourism. No less than 13 new tourist settlements have recently been built, for example, on the shores of Algeciras Bay, where there is now accommodation for 185 000 campers. Tourism has in fact become the key factor in Spain's expanding economy, ranking as the most important export industry ((3) Why export?) and accounting for 74% of the country's foreign earnings. These huge earnings, coupled with the trade stimulus given to builders and a host of workers in hotels, restaurants, shops, camps, transport firms and so on, help explain why Spain has the fastest economic growth rate in Europe.

The government underpins tourism by constructing state inns, hotels, roads, and by operating tourist offices in every large town in Spain and in 29 foreign countries. Such inducements help to explain the explosive increase in the number of foreign visitors to Spain from 1·9 million in 1953 to 19·1 million in 1968. The present aim is to attract at least 22 million tourists each year and to foster winter as well as summer tourism by building five new winter resorts.

THE BALEARIC ISLANDS (notably Majorca) are also sharing in the tourist boom. Until recently these islands were virtually unknown; now their names appear in brochures throughout Western Europe, and holiday-makers flock there every year. The Balearics attract because they are so typically Mediterranean. In addition to the certainty of a blazing summer sun they offer spectacular *karst* and coastal scenery. In Northern Majorca, for example, the rugged, garrigue-covered limestone of Sierra de Alfabia reaches well over 1500 m and falls precipitously to the sea.

For centuries the inhabitants of the Balearics have eked out a frugal existence from their peasant plots. Vines, oranges and other Mediterranean fruits, vegetables and cereals are grown, but water is scarce and irrigated *huertas* are found in only a few favoured spots. (4) Name these from the map below.

Some minerals are also worked and exported (*see map*). Nowadays, however, the tourist trade is fast becoming the main source of livelihood. Not only does it give employment in the hotel, catering and entertainment trades, but it also provides a market for traditional handicraft industries, especially jewellery, pottery, embroidery and basket making.

Salt evaporation pans — Lead Mines — IVIZA — FORMENTERA — 1580m — Palma — MAJORCA — Limestone & Lignite Quarries — MINORCA — HUERTAS — 0 30 km

The Balearic Islands

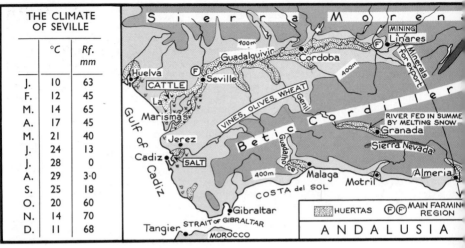

THE CLIMATE OF SEVILLE		
	°C	Rf. mm
J.	10	63
F.	12	45
M.	14	65
A.	17	45
M.	21	40
J.	24	13
J.	28	0
A.	29	3·0
S.	25	18
O.	20	60
N.	14	70
D.	11	68

THE SUNKEN LOWLANDS of Spain are located in the provinces of *Andalusia* in the south and *Aragon* in the north-east (*map p. 177*). In some ways these two basins are very similar: both were formed by the gradual collapse of land between fold mountain ranges and the ancient crystalline block of the Meseta; both are drained by important rivers; both contain deposits of fertile alluvium. (5) Name the fold ranges and the rivers referred to in the previous sentence.

The basins differ considerably, however, in climate. From the figures given above and on page 185, draw climate charts for Seville and Zaragoza. The lower latitude and altitude of Seville help to account for its distinctly warmer climate, but (6) how do you explain its much heavier rainfall? (*Hints: rain-bearing winds; relief.*) The hot, dry summers make irrigation essential for good crop yields in both regions, but Aragon is clearly the more arid. Farming there is difficult, and life is hard. In Andalusia, where water is rather more plentiful, farming is easier and the people more care-free. This is the Spain of the tourist posters, of opera and bull-fights, of fiestas and Holy Week processions.

For over 600 years Andalusia was a main centre of Moorish civilisation. The Moors entered Iberia from North Africa in A.D. 711 and soon overran the whole peninsula. They even crossed into France, where in 732 they were defeated in a great battle at the gates of Tours. The Moors then withdrew to Iberia, but it was not until 1492 that they were finally forced back across the Straits of Gibraltar. The long Moorish occupation left indelible marks on both landscape and people in Southern Spain; in particular they introduced the vital art of *irrigation* to this thirsty land.

The Court of Lions in the Moorish Palace of the Alhambra in Granada.

(7) Study the map carefully and complete the following:—

The River............ drains the lowlands of Andalusia, which are enclosed on the $\frac{south}{north}$ by the B.... C......... and on the north by the steep edge of the S..... M...... To the $\frac{north\text{-}east}{south\text{-}west}$ of S...... the river splits into many narrow distributaries and crosses a marshy area called L.. M....... The coastal lands of the Gulf of Cadiz are $\frac{rocky}{low\text{-}lying}$ and $\frac{marshy}{porous}$, and so are of $\frac{great}{little}$ agricultural value. Some $\frac{cattle}{sheep}$ are reared, mostly for the bull-rings, and $\frac{gypsum}{salt}$ is evaporated from coastal pans near C.... The northern lower slopes of the Betic Cordillera are planted with .... yards, ... groves and vast, unirrigated fields of ...... Elsewhere Southern Andalusia is too high, rugged and barren for cultivation, except in the *huerta* of the G.... valley near Granada, and on the coast near M....., M....., A...... and C........

(8) Answer the following questions:—

(a) Which two cities stand at each end of the main farming area?

(b) Suggest, with reasons, what kind of soils you would expect to find in the *huerta* districts.

(c) How does the Guadalquivir maintain its flow throughout the summer drought?

(d) Which Andalusian city has given its name to 'sherry'?

(e) Which port exports minerals obtained from the Sierra Morena?

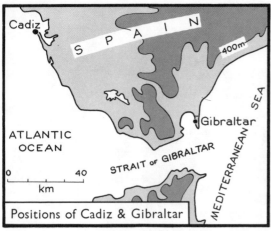

Positions of Cadiz & Gibraltar

Cadiz & Rota

(8) Describe the *site* of Cadiz and the nature of the harbour it commands. (9) Why was the *position* of Cadiz ideal for trading with both the 'Old World' and the 'New'?

THE MAIN TOWNS OF ANDALUSIA

Seville (480 000) is the largest city in Southern Spain and the main market, entertainment and commercial centre of Andalusia. It grew up at the head of navigation and lowest bridging point of the River Guadalquivir. In the early days of Spanish colonial conquest Seville enjoyed the monopoly of trade with Spanish America, and her merchants became very rich. In time, however, Seville's trading activities declined because the lower Guadalquivir silted up and ships became increasingly bigger. The port was saved from complete decay by the construction, in 1926, of the Tablada Canal. Seville exports agricultural produce (especially oranges) and minerals, and has food-processing and textile industries. Much of the city's income is now derived from the tourist trade.

Malaga (315 000) is the commercial capital of the Costa del Sol and the largest port of Andalusia. Agricultural produce from the Guadalquivir valley is sent to Malaga for export, via a tortuous railway across the Betic Cordillera. (10) Which river valley does this railway follow in its descent to the Mediterranean? (*Map p. 176.*) Malaga is best known for its exports of wine and raisins, but large quantities of fruit, early vegetables and olive oil are also shipped to Western Europe. The port has close trading links with Spanish Morocco.

Cordoba (219 000) is an important market town, where agricultural produce from the nearby *huertas* is assembled, processed and despatched to the coast for export. Recently the town has acquired an electrical industry, and it is also an important tourist centre. Its partially ruined mosque, second only in size to the great Kaaba of Mecca, is one of the world's architectural wonders.

Cadiz (130 000) succeeded Seville as the main port trading with Spanish America, and was for a while the chief port of Spain. Modern Cadiz is mainly concerned with the export of sherry and olive oil. It is also a naval base and the chief port for trade with the Canary Islands. On the mainland opposite Cadiz lie the arsenals and shipyards of Matagorda, San Fernando and La Carraca, one of the biggest concentrations of heavy industry in Spain.

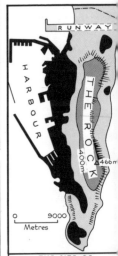

THE SITE OF GIBRALTAR

Gibraltar is a British colony. It was seized in 1704 and a naval base was established at the foot of the famous limestone Rock. The site of the naval dockyards and other installations is shown on the map; they still partly support a resident population of about 25 000, but tourism is rapidly gaining in importance. In addition about 5000 Spaniards cross the border each day to work in Gibraltar. (11) Describe (*a*) the advantage of Gibraltar's site and (*b*) the strategic importance of its position (*map, above*).

wheat
tobacco
grapes
olives
groundnuts
almonds
cotton
rice
dates
citrus fruits
sugar cane
vegetables

The list on the left shows some of the great variety of temperate and sub-tropical crops grown in the *huertas* of Andalusia. Total output is smaller than in Murcia and Valencia, for irrigated land is comparatively scarce, but cargoes of fruit, vegetables, wines, olive oil and processed foods are regularly exported to Western Europe. Most of the larger towns have industries based on local agricultural products; details are given in the notes opposite.

Aragon (the Ebro Basin) is isolated from the rest of Spain and from the sea. ((*12*) *How?*) As well as being remote, the Basin is one of the driest parts of Spain. (*13*) State (*a*) the total mean annual rainfall for Zaragoza and (*b*) its rainiest season(s).

Isolation and aridity are reflected in the economic backwardness of much of the region. Vast areas consist of dry mountain steppe, suitable only for the summer pasturing of sheep. Cultivation is limited by water shortage and by the widespread occurrence of salt deposits in the top-soil. (*14*) Suggest the origin of these deposits. (*Hint: evaporation.*) Water for irrigation is obtained mainly from the Ebro and its Pyrenean tributaries. (*15*) Describe, from the map below, the location of the main irrigated districts and canals. The main crops in this region are wheat and sugar beet, but vines, almonds, stone-fruit and olives are also grown.

The only large city is Zaragoza (362 000), where in addition to food-processing industries there are engineering firms making vehicles, electrical goods, railway equipment and machine tools. (*16*) Suggest why Zaragoza is the railway hub of Northern Spain.

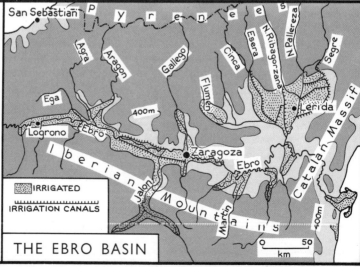

THE CLIMATE OF ZARAGOZA		
	°C	Rf. mm
J.	6	18
F.	8	20
M.	11	23
A.	13	28
M.	18	38
J.	21	28
J.	24	15
A.	24	15
S.	21	23
O.	14	33
N.	9	30
D.	6	20

IRRIGATED
IRRIGATION CANALS

THE EBRO BASIN

0   50
km

of Spain consist of Galicia, Asturias and the Basque Provinces. These lands are isolated from the rest of the country, and from one another, by difficult mountainous terrain. Thus they have tended to develop separately. (*17*) How is this reflected in the pattern of main railways?

Even so the coastlands have geographical similarities in their mountain relief, oceanic outlook and ' Northwest European ' climate (*pp. 14–18*), which is similar to that of South-west England though much warmer and sunnier. Rain falls in all months, but summers are hot and comparatively dry and irrigation, though not vital, is sometimes used to obtain maximum crop yields.

Northern Spain is at present experiencing a trade ' boom ', and Asturias, with 9/10 of the country's bituminous coal, 3/5 of the iron ore reserves and a large portion of the readily available water, is obviously a key economic region. The other main industrial bases in Spain are Barcelona (*p.* 180) and Madrid (*see overleaf*).

THE NORTHERN COASTLANDS

Galicia resembles Cornwall and Brittany, for (*a*) the granite uplands of the interior are barren and virtually useless, and farming is confined to the river valleys and coastal lowlands; and (*b*) a 'drowned' coast, with many fine, deep, sheltered rias, favoured the growth of seafaring and fishing. Sardines and tunny are caught off the Spanish coast. Galician fishermen also catch skate off Ireland, and cross the Atlantic to the cod fisheries of the Grand Banks. The main fishing ports are underlined on the map. *Ferrol* and *Corunna* are also naval bases.

Most Galicians live by subsistence farming on small peasant plots. It is hot enough for maize to be the main summer crop, and wet enough for cows to be kept for milk and as draught animals. Oats, barley, cabbages and roots help vary the diet, but living standards are low and there is a long history of emigration. Galicia has no fuels, few minerals and little industry apart from food-processing, boat-building and fertiliser production (at a new government-built plant at *Puentes*). Galicia produces about 10% of Spain's h.e.p. (*18*) Suggest why Galician ports, although capable of taking the largest ocean-going vessels, have never grown into flourishing centres of commerce and industry.

Asturias is important for its minerals. Two-thirds of Spain's bituminous coal and nearly all its anthracite come from the coalfield east of *Oviedo*. This is one of the very few places in Spain with old-established heavy industries, notably iron and steel at Oviedo and chemicals and glass at *Gijon* and Oviedo. At *Avilès* a huge new steel plant employs 17,000 men and has an annual output of 900 000 tonnes, i.e. two-thirds of Spain's total steel production.*

* The latter is still comparatively small—about the same as that from one large steel works in South Wales.

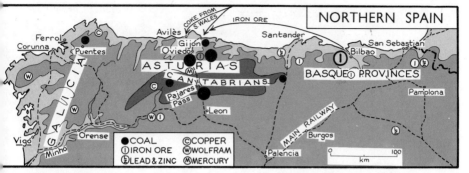

The Asturian steelworks use some local coking-coal and a little impure iron ore mined near Oviedo, but the bulk of these raw materials are imported: from the map (*19*) name (*a*) the sources of these imports and (*b*) the ports involved. (*20*) Name and locate the sources of other minerals produced in Asturias. Development is hindered by poor communications. (*21*) Name the pass by which the one main road and one main railway to interior Spain cross the Cantabrians.

The Basque Provinces are a major centre of Spanish industry. This region had the advantages of large supplies of high-grade iron ore, good natural harbours and adequate communications with both Spain and France. In addition the Basque people, who differ in race, language and temperament from other Spaniards, are enterprising and energetic. In the late 19th century the old-established Basque metal-working and woodcraft industries were greatly developed. *Bilbao* and *Santander* built up substantial iron and steel and engineering industries using local iron ore and coal imported from South Wales. The latter was received in exchange for iron ore. Today this trade pattern has changed; ore exports to South Wales have dwindled and the Basque steelworks rely on Asturias and the U.S.A. for their coal supplies. In addition to steel, Bilbao (320 000) has chemical, glass and fishing industries and is the region's commercial and industrial capital. The Basque coast is a very popular holiday district and *San Sebastian*, in addition to paper, cement, food-processing and textile industries, is a major European resort.

Big efforts are being made to develop and modernize the Spanish iron and steel industry, but the latest methods for making steel require enormous quantities of water. This is a serious hindrance in a country which mostly suffers a drastic drought for much of the year. A glance at the lower map on page 178 will show why seven of the eight major integrated steelworks so far built in Spain are located in Asturias and the Basque Provinces: the eighth is at Sagunto near Valencia. Problems of water supply elsewhere therefore reinforce the old-established tendency for steelworks to be located in the north and north-west. Furthermore an increasing dependence on foreign supplies of iron ore and coking coal ensures that new modern plants are built on the coast, i.e. on **tide-water sites,** preferably where the enormous modern bulk carriers can find deep, sheltered anchorage for unloading. Other examples of this trend include Europort, IJmuiden, Taranto and Lisbon (see pp. 114, 116, 171 and 192).

THE MESETA is the largest but economically the least productive region of Spain. It consists of vast, arid plateaus and high Sierras, trenched by a few deep river valleys. The dry, infertile soils of the Meseta make cultivation difficult, and settlements are few and far between. Quite recently a traveller wrote:—

> "These bare, empty plains are frightening in their stark immensity. For mile after mile we plodded northwards over the dusty, yellow earth, never a tree, a house or a man to be seen. At times we dismounted to lead the horses by narrow, steep paths across rock-strewn, waterless gullies . . . for seven hours we rode northwards through the solitude and burning heat, but the gleaming snows of the distant Sierra looked as remote as ever. Then, quite suddenly, we saw the village, its white walls shimmering against a barren hillside."

MAIN TOWNS OF THE MESETA

Madrid (2 610 000) has only one natural advantage as the capital of Spain—centrality. It was chiefly for this reason that Philip II chose the city as his capital and built a massive granite palace near by. Until the present century Madrid had no significant industries, for there were no important local supplies of raw materials or minerals and no adequate communications. Today main roads and railways converge upon Madrid from all parts of Iberia (see *map*). Improved accessibility has favoured the growth of industries ((22) Why?), and during the past two decades Madrid has experienced a minor 'boom', attracting labour from all over Spain. Madrid's expanding industries include the manufacture of vehicles and electrical equipment, as well as of consumer goods such as clothing, foodstuffs and furniture. H.e.p. and water are obtained from the rivers of the upper Tagus Basin (see *map*). Politics and administration, however, remain the main functions of Madrid, and most of its imposing new buildings are government offices.

Leon, Burgos and *Valladolid* are small provincial capitals, market centres and minor route foci.

Toledo is one of the oldest cities of Iberia. It grew up on a rocky peninsula, a magnificent defensive site guarding a bridging point on the Tagus. Occupied in turn by Romans, Visigoths, Moors and Crusaders, it gained fame in medieval times as a centre of learning and culture. When Madrid was chosen as the capital of a united Spain, Toledo gradually fell into decay. Its inhabitants now live chiefly on the tourist trade.

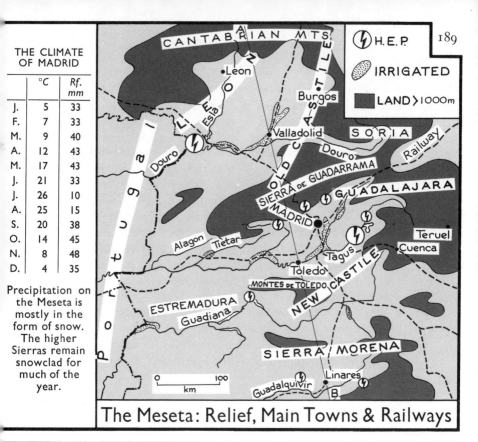

THE CLIMATE OF MADRID

	°C	Rf. mm
J.	5	33
F.	7	33
M.	9	40
A.	12	43
M.	17	43
J.	21	33
J.	26	10
A.	25	15
S.	20	38
O.	14	45
N.	8	48
D.	4	35

Precipitation on the Meseta is mostly in the form of snow. The higher Sierras remain snowclad for much of the year.

The Meseta: Relief, Main Towns & Railways

The diagram represents a cross-section along line AB on the map. (23) Copy the diagram, naming (a) the towns T1 and T2 (b) the rivers R1–R3 and (c) the mountainous areas numbered 1–4.

The Meseta is divided into two parts by the high Sierra de Guadarrama. South lies the 'basin' of New Castile and Estremadura, and north the 'basin' of Old Castile and Leon.

In both, the climate is arid and extreme. Using the figures given above (24) make a climate chart for Madrid. (25) Compare with that for Seville (p. 182) and explain the differences in the mean annual temperature ranges of the two places. (Hints:— altitude; distance from the sea.)

Where irrigation is impossible wheat and barley are grown every second or third year in rotation with beans and fodder crops. In better watered districts, such as the major river valleys and the upper Tagus Basin, good crops of cereals, grass, vines, olives and sugar beet are obtained, but true huertas are rare. (26) On which rivers are there reservoirs for irrigation and h.e.p. generation?

Very large areas are useful only as scanty pasture for sheep and goats. In summer over a million animals browse on the high pastures of Soria, Guadalajara and Teruel; in winter they are taken by rail or lorry to lower ground in Estremadura.

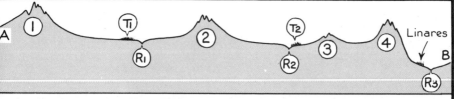

DIAGRAMMATIC CROSS-SECTION ALONG THE LINE AB ON THE MAP ABOVE

(2) PORTUGAL

Like Spain, Portugal is economically backward. There is little heavy industry, and most of its 9 600 000 people live on the land, either as peasant farmers or as labourers on big estates. Farming methods are primitive, 40% of the population can neither read nor write, farm holdings are often minute and there is much seasonal unemployment. Not surprisingly, therefore, the standard of living is one of the lowest in Europe.

The country may be divided into two main regions, (i) *THE UPLANDS*, which form the western margin of the Meseta and (ii) *THE COAST LOWLANDS*. The Uplands are similar to those of interior Spain and are very sparsely inhabited. Their southern parts are covered with scrub and poor pasture, but in the north, where the rainfall is heavier, there are cork-oak forests. The only important products of the Uplands are cork and some minerals. (*27*) Name and locate these minerals, and state the ports through which they are exported.

The Lowlands vary considerably because of differences in rainfall and soils, and are divided into two climatic sub-regions by the Sierra da Estrela. (*28*) Where does this range reach the coast? (*29*) Use the map on page 178, and the map and details opposite to complete the following sentence:—

To the $\frac{north}{south}$, in the Douro plains, rainfall is $\frac{scanty}{plentiful}$. All places receive at least ... millimetres of rainfall and there are $\frac{only\ two}{no}$ rainless months. South of the S. da Estrela the rainfall $\frac{increases}{decreases}$ rapidly: Lisbon receives ... millimetres, but the south coast only ... millimetres, with ... months of almost complete drought.

(*30*) How do you account for the differences in the rainfall of the northern and southern parts of the Portuguese lowlands? (*Hints: 'Westerlies' and latitude: rain-shadow.*)

(*31*) Use the bar diagram to work out the percentage of Portugal which is (*a*) *cultivated*, (*b*) *under pasture*, (*c*) *forested*, (*d*) *mountain and heath*. (*32*) Use the details given on the map to write notes on farming in Portugal.

Lisbon (8 320 000) is the capital of Portugal and the industrial and commercial centre of a still considerable empire. ((*33*) *Name some overseas territories of Portugal.*) It is also by far the largest Portuguese city and port. (*34*) Suggest natural advantages of site

Cultivated	Pasture	Forested	Mountain & Heath

PORTUGAL: LAND USE

Rainfall (mm)	J	F	M	A	M	J	J	A	S	O	N	D
OPORTO	160	113	148	85	88	40	20	25	50	105	148	168
LISBON	88	85	88	73	50	20	5	5	38	78	113	103
FARO	70	53	73	30	20	5	0	0	18	50	65	68

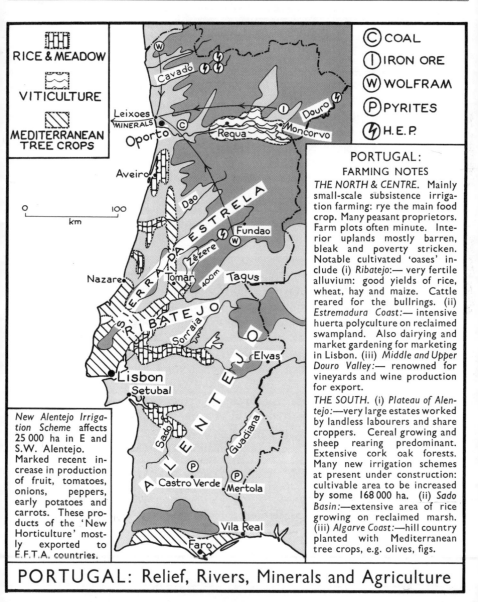

RICE & MEADOW

VITICULTURE

MEDITERRANEAN TREE CROPS

© COAL
Ⓘ IRON ORE
Ⓦ WOLFRAM
Ⓟ PYRITES
⚡ H.E.P.

New Alentejo Irrigation Scheme affects 25 000 ha in E and S.W. Alentejo. Marked recent increase in production of fruit, tomatoes, onions, peppers, early potatoes and carrots. These products of the 'New Horticulture' mostly exported to E.F.T.A. countries.

PORTUGAL: FARMING NOTES

THE NORTH & CENTRE. Mainly small-scale subsistence irrigation farming: rye the main food crop. Many peasant proprietors. Farm plots often minute. Interior uplands mostly barren, bleak and poverty stricken. Notable cultivated 'oases' include (i) *Ribatejo:*— very fertile alluvium: good yields of rice, wheat, hay and maize. Cattle reared for the bullrings. (ii) *Estremadura Coast:*— intensive huerta polyculture on reclaimed swampland. Also dairying and market gardening for marketing in Lisbon. (iii) *Middle and Upper Douro Valley:*— renowned for vineyards and wine production for export.

THE SOUTH. (i) *Plateau of Alentejo:*—very large estates worked by landless labourers and share croppers. Cereal growing and sheep rearing predominant. Extensive cork oak forests. Many new irrigation schemes at present under construction: cultivable area to be increased by some 168 000 ha. (ii) *Sado Basin:*—extensive area of rice growing on reclaimed marsh. (iii) *Algarve Coast:*—hill country planted with Mediterranean tree crops, e.g. olives, figs.

PORTUGAL: Relief, Rivers, Minerals and Agriculture

Some Important Industries of LISBON
Olive-oil processing
Oil-seed crushing
Jute manufacture
Cable making
Shipbuilding and repairing
Cotton manufacturing
Wine making

Site and Position of Lisbon

A new iron and steelmill at Seixal, which yields more than I million tonnes of pig iron per annum, is scheduled to play an important role in the industrial development of Portugal. Typical steel products include pipelines, reinforcing rods for concrete, wire for fencing and nails. Portugal has no coking coal and so the steelworks had to be located at a coastal site to receive imported coke, mainly from the U.S.A. (See also note, p. 187, on tidewater industrial location). The site at Seixal also has the advantage of a safe, deep water anchorage for bulk carriers, water for cooling from the River Tagus and from local wells, plenty of cheap labour in the vicinity of Lisbon, and hydro-electricity from the Tagus development schemes (photo overleaf). Iron ore from the Guadiana valley is brought to Seixal by ship and limestone is obtained from nearby Arrábida. At present production costs in the steelmill are high, and so the enterprise is protected from foreign competition by high import tariffs. (35) Do you think that this arrangement is a real advantage to Portugal?

and position which favoured its growth. (36) Which of the industries listed above reflect Lisbon's long history of trade with Portuguese colonial territories? Which are typical 'Mediterranean' food-processing industries?

Lisbon's merchants conduct an entrepôt trade in such colonial goods as oil-seeds, cotton and hides, and export cork, wine and minerals from the port's immediate hinterland. In addition Lisbon is (i) a N.A.T.O. naval base, (ii) a port-of-call for Italian liners on the Atlantic routes, (iii) a base for fishing vessels plying to the Grand Banks cod-fisheries, and (iv) a terminal on both North and South Atlantic air routes. (37) How does Lisbon's geographical position favour each of these functions?

Oporto (326 000) is chiefly concerned with the wine trade, especially with the export of 'port' wine to England. In recent decades, however, manufacturing industries have developed rapidly and it is now Portugal's second industrial centre. The main industries are woollen, rayon and silk textiles, fish canning and the manufacture of textile machinery and electrical equipment.

Setubal is the principal centre of Portuguese sardine fishing and canning. Vila Real is the chief tunny-fishing port. Between April and July the tunny are netted as they enter the coastal lagoons to spawn, and are then slaughtered with spears.

PORTUGAL'S FOREIGN TRADE

Main Imports	Main Exports
Foodstuffs, notably dried cod, cereals, coffee and sugar.	Foodstuffs, notably sardines, wines and olive oil.
Raw Materials, notably iron and steel ingots, raw cotton, coal and fertilisers.	Raw Materials, notably resin and turpentine, wolfram, pyrites and pit-props.
Manufactures, notably vehicles, iron and steel goods, and dyestuffs.	Note the dependence on exports of agricultural, forest, fish and mineral products.

(Above): *Collecting bark from cork-oak trees near Faro. Note the primitive form of transport still characteristic of rural Portugal. Portugal is the world's leading producer and exporter of cork, most of which comes from the oak forests of western Alentejo. Uses of cork include bottling, gaskets, fire-proofing and lifebelts. No entirely satisfactory synthetic substitute has yet been discovered.*

(Below): *Grafting vines in the Douro Valley. Only expert farming can wring a living from such arid and stony ground.*

New dam, irrigation reservoir and h.e.p. station on the R. Tagus, Portugal. Note the tree crops on the hill slopes.

Modern Economic Development in Iberia is symbolised by this large new h.e.p. station. Similar large projects have recently been completed, or are under construction, on other headstreams of the Tagus as well as on the Douro and Ebro. Altogether there are now well over a thousand h.e.p. stations in Iberia, though many of them are small-scale. In addition several thermal-electric power stations have recently been built to utilise small deposits of lignite, e.g. at Escatron in the Ebrox Valley. Spain also has a huge new petroleum refinery at Escombreras. The emphasis that both the Spanish and Portuguese governments are laying on increased electric power production reveals the prime weaknesses of both countries' economies, i.e. lack of fuel and power, without which any industrial expansion is seriously handicapped.

Even so, both Spain and Portugal are making big efforts to improve industrial outputs through Development Plans backed by heavy U.S. investment. In Portugal the expansion is mainly in engineering, chemicals, food processing, forestry (*see p. 193*), textiles (*aided by E.F.T.A. sales*) and fishing. A thriving fishing industry is vital for Portugal's future prosperity, for fish and fish products form the largest group of exports. The harvest of tunny fish off the Azores, Canaries, S.W. Africa and mainland Porgutal is doubling every two years and there are also very large increases in catches of Newfoundland cod and W. African pargo and mullet. In addition the Portuguese Antarctic whaling fleet brings home more than 400 whales every year.

In Spain the immediate aim is to create an extra million new jobs. Many of them are to be in service industries such as tourism (*details p. 181*), but steel production is rapidly approaching a

A farmer in Tarragona breaking up the soil with a digging stick.

target of 8·5 million tonnes per annum and there is a boom in the output of ships, chemicals, cement, vehicles and refined petroleum products.

Iberian crop yields are amongst the lowest in Europe but efforts now being made to improve farm outputs include:

increased use of fertilizers and farm machinery;
enforced consolidation of very small farm holdings;
confiscation by the State of badly farmed estates;
expansion of areas under irrigation (*the largest schemes are in Spain in Badajoz Province on the lower Guadiana, in parts of Aragon and in Jaen Province*);
re-afforestation of hill-slopes, to reduce the run-off of rainwater and prevent further soil erosion.

Some of the more important *recent* industrial developments in Iberia are listed here. (*38*) Show this information by means of black dots and accompanying labels on a *large* outline map (the location of every place in the list is given on the maps in this chapter). Then add to your map two large *red* dots to represent the main industrial areas of (i) Barcelona and (ii) the Basque Provinces, and add labels to show the main established industries of these areas. Give your completed map an appropriate title.

TOWN	PRODUCT	TOWN	PRODUCT
Avilès	Steel	Madrid	Vehicles,
Barcelona	Vehicles		Electrical
Cartagena	Oil Refinery,		equipment
	Ships,	Oporto	Engineering,
	Chemicals		Chemicals
Cordoba	Electrical	Oviedo	Steel
	equipment	Santander	Chemicals
Ferrol	Ships	Burgos ⎱	Nuclear
Galicia	Chemicals	Guadalajara ⎰	Power
San Sebastian	Rubber		
Lisbon	Engineering		
	Chemicals,		
	Steel		

(3) GREECE

Modern Greeks are in part the direct descendants of the illustrious, prosperous and highly civilised people who once controlled the greatest maritime power of the ancient world. Yet today Greece is a rather backward country, a land of simple and often poverty-stricken peasant farmers, 90% of whom work on farm-holdings of less than 5 hectares. The more obvious reasons for the relative decadence of modern Greece are listed below. They are a reminder of the tangled web of history and geography which influences the destiny of all peoples. (39) Which of these reasons are — historical? — geographical? — a mixture of both?

(i) Before the modern Greek state was created in 1830 the country had endured nearly 400 years of tyrannical and inefficient rule by the Turks, during which many ancient irrigation works and drainage systems fell into disrepair.

(ii) Greece suffers from a severe summer drought, made worse by the widespread occurrence of infertile, permeable limestone soils.

(iii) The low-lying basins, valleys and small coast plains are swampy, and so poor drainage hampers cultivation of the better land. The swamps also have a dismal record as malarial plague-spots.

(iv) The 16th-century 'Age of Discovery' drew merchants and traders away from the Mediterranean, the eastern part of which became virtually a backwater until the Suez Canal was built.

Agriculture produces a living for one half of all Greeks, despite the fact that only 28% of the total land area is cultivable. Peasant farmers in Greece face seemingly endless difficulties:

"On a thin soil, patiently husbanded by years of careful work, the people produce the crops that are their livelihood. Every stretch of flat land is used, and where the mountain slopes become too steep to hold the soil they are terraced to catch the eroded earth that the rain brings down. Fires, woodcutting and goats have diminished the forests which once covered much of the land, and the autumn and winter rains cause flash floods on ground empty of all but a thin scrub.

"In the alluvial plains fed by permanent rivers, oranges, cotton and even rice can grow and flourish. But in the mountains almond and Spanish chestnut, olive and fig are the familiar trees. A man can live as long as he has olives, cheese from his sheep or goats and wine from his grapes. The olive is the true wealth of the land . . . and is grown as high in the mountains and as far north as it will survive. . . . The vine thrives anywhere, taking hold in the poorest soil, and few country families are without their small vineyard or vine trellis before their house. The grain crops are harder to produce . . . grown to be harvested in May or June before the sun burns the landscape dry and bare. But for every crop that is produced, skill and backbreaking

GREECE

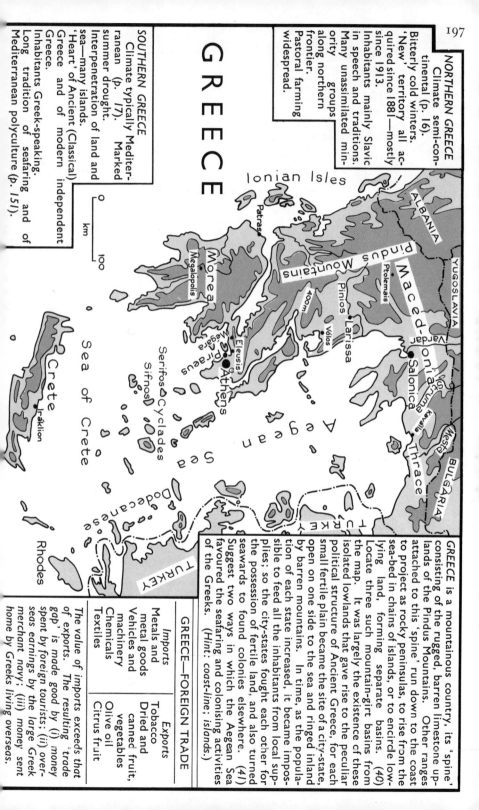

NORTHERN GREECE

Climate semi-continental (p. 16).
Bitterly cold winters. 'New' territory all acquired since 1881—mostly since 1913. Inhabitants mainly Slavic in speech and traditions. Many unassimilated minority groups along northern frontier. Pastoral farming widespread.

SOUTHERN GREECE

Climate typically Mediterranean (p. 17). Marked summer drought. Interpenetration of land and sea—many islands. 'Heart' of Ancient (Classical) Greece and of modern independent Greece. Inhabitants Greek-speaking. Long tradition of seafaring and of Mediterranean polyculture (p. 151).

GREECE is a mountainous country, its 'spine' consisting of the rugged, barren limestone uplands of the Pindus Mountains. Other ranges attached to this 'spine' run down to the coast to project as rocky peninsulas, to rise from the sea-bed in chains of islands, or to encircle low-lying land, forming separate basins. (40) Locate three such mountain-girt basins from the map.

It was largely the existence of these isolated lowlands that gave rise to the peculiar political structure of Ancient Greece, for each small fertile plain became the site of a city-state, open on one side to the sea and ringed inland by barren mountains. In time, as the population of each state increased, it became impossible to feed all the inhabitants from local supplies; so the city-states fought each other for the possession of fertile land, and also turned seawards to found colonies elsewhere. (41) Suggest two ways in which the Aegean Sea favoured the seafaring and colonising activities of the Greeks. (*Hint: coast-line; islands.*)

GREECE—FOREIGN TRADE

Imports	Exports
Metals and metal goods	Tobacco
Vehicles and machinery	Dried and canned fruit, vegetables
Chemicals	Olive oil
Textiles	Citrus fruit

The value of imports exceeds that of exports. The resulting 'trade gap' is made good by (i) money spent by foreign tourists; (ii) overseas earnings by the large Greek merchant navy; (iii) money sent home by Greeks living overseas.

Preparing ground for tobacco plants in the Nestos Valley. (See also notes p. 24 bottom right.)

labour is required. No modern tools can work in ground that is little better than broken rock. The hoe and a primitive iron-shod plough which does no more than scratch the land are the main implements on ground that would break a share and throw a tractor. Threshing is still often done by the trampling of horses' hooves or by wooden sleds shod with sharp stones dragged over the corn; and winnowing is performed with huge wooden flails."*

Pastoral farmers have an even harsher existence. In the Macedonian Mountains, for example:

" . . . shepherds will spend the summer months in their *kalyvas*—rough huts of branch and bracken—far from human habitation, alone with their sheep and wolf-like dogs."*

The Greek Government has recently made great efforts to improve conditions, for farm products make up the bulk of the country's exports. Irrigation and reclamation schemes have considerably increased the cultivable area, notably in parts of 'New' Greece.

THE GREEK ISLANDS are renowned for their beauty and for their ancient history. Beyond this, however, the islands have few important resources and only 15% of the Greek people live there. As in mainland Greece the islanders are mainly peasant farmers. Even on Crete, the largest island, cultivation is severely limited by the mountainous terrain, limestone soils and summer drought. Olives, wheat and vines are grown and some wine is exported, notably Malmsey from the Cyclades.

* Geoffrey Chandler, 'Country Life in Greece', *Geographical Magazine.*

INDUSTRY in Greece has long been hampered by a lack of home-produced fuels and raw materials and by a small home market. In 1962 Greece became an associate member of the Common Market with access to 185 million potential consumers, but industrial growth is still slow. Many towns have small traditional handicraft industries such as glove and lace making, but there are only two sizeable manufacturing centres, Athens and Salonika.

Athens (1 853 000) grew up in classical times round the fortress of the Acropolis, built on a small ridge in a fertile coastal plain. It was partly because of the prestige attached to its name that Athens was chosen as the capital of modern Greece. Another reason was its geographical position: (42) how has this favoured the city's growth as a metropolitan centre? Since 1830 Athens has grown rapidly, especially in the last few decades. Now it has merged with its port, Piraeus, to form a built-up area covering most of the plain. Athens itself is mainly an administrative and tourist centre; many sightseers and scholars visit the city to inspect the remains of its ancient buildings.

Piraeus is the main port of Greece and the principal industrial centre. Its main industries, *food-processing, chemicals, textiles, shipbuilding and marine engineering* are based on (a) agricultural raw materials (mainly home produced) and (b) general maritime activities. Piraeus is the centre of the world's largest merchant fleet (26·5 million tonnes), but though the ships are Greek-owned less than half of them actually sail under the Greek flag. ((43) Why?). This fleet is the country's largest single source of foreign currency. ((44) *Explain this*). Fifteen kilometres north of Piraeus at the port of *Eleusis* is the only Greek iron and steel manufacturing centre, where two blast furnaces, a steelworks and rolling and

drawing mills give employment to 2000 men. Iron ore and coke are imported.

Salonika (378 000) is the growing metropolis of ' New ' Greece, and the only large industrial centre outside the Piraeus. Its industries resemble those of Piraeus. The port includes a ' Yugoslav Free Zone ' because its hinterland includes part of Yugoslavia as well as all Eastern Greece. (*Map, p. 197*).

The present Greek 5-Year Plan aims to expand steel-making up to a target of 1 million tonnes p.a. and to increase the output of shipyards, chemical works, aluminium and other metal-working industries. Power supplies are to be expanded by (i) modernization of the oil refinery at Megára, (ii) the building of power stations to be fuelled by lignite quarried at Megalopolis and Ptolemais and (iii) the construction of more hydro-electric power stations on eleven Greek rivers: at present only one-third of the country's hydro-electric potential is tapped. By 1974 the lignite and hydro-power stations will supply 61% of the country's electricity, the remainder coming from oil-fired stations.

New industries, e.g. food processing and light engineering are also being established at Vólos, Patras, Iráklion and Kavalla (*map p. 197*), but the country's industrial plans are being seriously hampered by an annual emigration of tens of thousands of workmen: nearly 200 000 Greeks now work permanently in Common Market countries. The reason is the low wages in Greece compared with those in most other West European states. In spite of determined efforts to industrialize only 770 000 of the 3·7 million workers in Greece are as yet in industrial jobs. (45) Make notes on the similarities of these problems with those afflicting Southern Italy (*see pages 168–173*).

Mining is locally important, e.g. for iron ore on Serifos and Sifnos, and miners' wages help lift living standards above subsistence level. So, too, does the money spent by foreign tourists on accommodation, entertainment and locally produced souvenirs. The Ionian Isles are more prosperous than most other Greek islands, mainly because throughout history they have been less affected by invasions, piracy and plunder. The Isles are well known for their currants and olives.

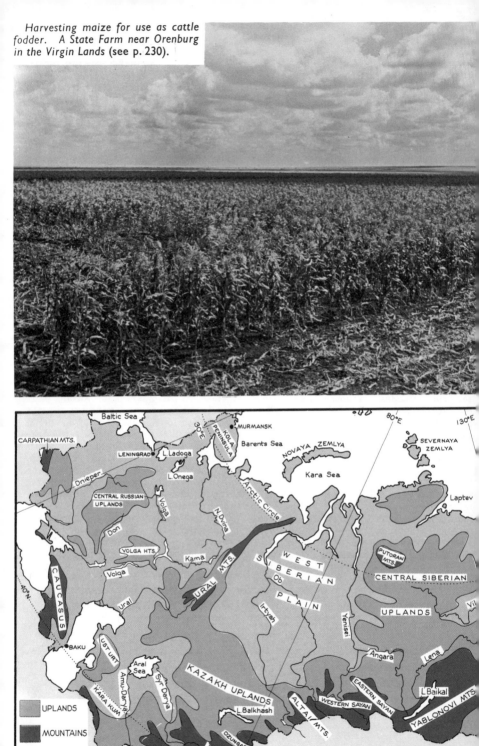

Harvesting maize for use as cattle fodder. A State Farm near Orenburg in the Virgin Lands (see p. 230).

U.S.S.R.: General Landform

UPLANDS

MOUNTAINS

0 — 500 km

Baltic Sea
CARPATHIAN MTS.
Barents Sea
MURMANSK
KOLA PENINSULA
LENINGRAD
L. Ladoga
L. Onega
NOVAYA ZEMLYA
Kara Sea
SEVERNAYA ZEMLYA
Dnieper
Arctic Circle
CENTRAL RUSSIAN UPLANDS
Volga
N. Dvina
Don
Kama
Laptev
VOLGA HTS.
URAL MTS.
WEST SIBERIAN PLAIN
Ob
PUTORAN MTS.
CENTRAL SIBERIAN
Volga
Ural
Irtysh
UPLANDS
CAUCASUS
Yenisei
Vil
BAKU
UST URT
Aral Sea
Amu-Darya
Syr-Darya
KAZAKH UPLANDS
Angara
Lena
L. Baikal
KARA KUM
L. Balkhash
ALTAI MTS.
WESTERN SAYAN
EASTERN SAYAN
YABLONOVI MTS.
DZUNGARIAN GATE
TIEN-SHAN
PAMIRS
30°E
80°E
130°E
40°N

CHAPTER 9

The Union of Soviet Socialist Republics (U.S.S.R.)

THE U.S.S.R. is the largest country in the world, covering one-sixth of the land surface of the globe. (*1*) How far is it (*a*) from Murmansk to Baku and (*b*) from Leningrad to Vladivostok? Train journeys between the two latter towns take five days. (*2*) How many degrees of longitude are crossed *en route*?

This immense territory contains varied land forms, including desert, tundra and mountain ranges—all on a huge scale—but the most typical scenery consists of gently undulating lowlands covered with taiga or steppe. The photograph shows part of the plain which runs almost without interruption from the western borders of the U.S.S.R. to the plateaus of Central Siberia. (*3*) Name the fold mountains which cross this plain from north to south.

201

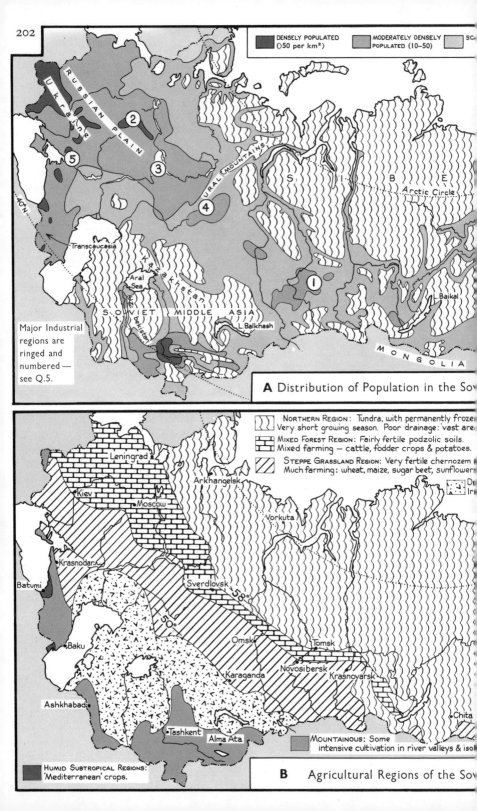

A Distribution of Population in the So[viet Union]

Major Industrial regions are ringed and numbered — see Q.5.

DENSELY POPULATED (>50 per km²)   MODERATELY DENSELY POPULATED (10–50)   SC[ARCELY POPULATED]

RUSSIAN PLAIN — Ukraine — URAL MOUNTAINS — Arctic Circle — S   B   E — L.Baikal — MONGOLIA — Transcaucasia — Aral Sea — Kazakhstan — S O V I E T   M I D D L E   A S I A — Uzbekistan — L.Balkhash — 40°N

B Agricultural Regions of the So[viet Union]

NORTHERN REGION: Tundra, with permanently froze[n subsoil]. Very short growing season. Poor drainage: vast are[a]

MIXED FOREST REGION: Fairly fertile podzolic soils. Mixed farming — cattle, fodder crops & potatoes.

STEPPE GRASSLAND REGION: Very fertile chernozem [soils]. Much farming: wheat, maize, sugar beet, sunflowers

De[sert] Ir[rigation]

MOUNTAINOUS: Some intensive cultivation in river valleys & iso[lated]

HUMID SUBTROPICAL REGIONS: 'Mediterranean' crops.

Leningrad — Kiev — Moscow — Arkhangelsk — Vorkuta — Krasnodar — Batumi — Sverdlovsk — 58° — 50° — Baku — Omsk — Tomsk — Novosibersk — Krasnoyarsk — Karaganda — Ashkhabad — Tashkent — Alma Ata — Chita

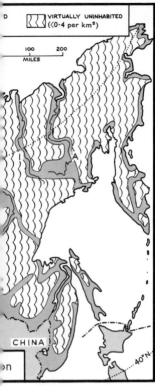

VIRTUALLY UNINHABITED
(<0·4 per km²)

100 200
MILES

CHINA

Taiga, with thin, infertile soils.
in summer. Little agriculture.

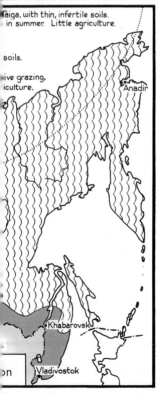

soils.

sive grazing,
iculture.

Anadir

Khabarovsk

Vladivostok

Note from Map **A** that most people in the U.S.S.R. live in the western part of this plain, but that there are also clusters and ribbons of dense population elsewhere. The continuously settled area is shaped like a wedge, tapering westwards into Western Siberia from a broad base against the country's western borders.

This pattern of settlement has evolved mainly because this part of the U.S.S.R. is the most suitable for agriculture. Although the U.S.S.R. is a major industrial power, 48% of her total population of 242 million still live in the country. (4) State (a) the main types of soils found within the principal inhabited region, (b) the geographical factors which hinder farming (i) to the north of approx. 58°N. and (ii) in Caucasia and Soviet Middle Asia (i.e, south of approx. 50°N.).

The major industrial regions also lie within the 'wedge'. They are:—the Moscow region, the Eastern Ukraine (Donbas), the Urals, the Volga Valley (Povolzhye) and the Kuznetsk Basin. Each of these regions is ringed and numbered on Map **A**; with the help of an atlas (5) give each region its appropriate name. Reasons for the development of these regions are referred to later in this chapter. The dense population of the Western Ukraine consists largely of farmers.

Notice the isolated areas of denser population found (i) in Transacausia and (ii) in parts of Middle Asia. In such areas the original settlement was usually because the land and climate favoured agriculture. The summers are long and hot and there is plenty of water from the adjoining mountains for large-scale irrigation schemes. (6) Name the two large rivers which drain to the Aral Sea. Uzbekistan and southern Kazakhstan have been developed as the principal cotton-producing lands of the U.S.S.R., whilst the western part of Transcausia has a Mediterranean climate and is one of the few parts of the U.S.S.R. in which crops such as citrus fruit can be grown. In recent years the mineral wealth of the Caucasus and Middle Asia has been exploited on an ever-increasing scale, with a corresponding increase in population (details, pp. 226-229). There are also isolated spots of settlement in the Far East, including fishing ports on the Pacific coast and farming and forestry communities in intermontane basins, mostly adjoining the Chinese frontier.

Look again at the map: vast parts of the U.S.S.R. are as yet uninhabited. There are a few remote and isolated mining camps in the tundra wastes and mountain ranges of Siberia and the Far East, but it seems likely that the severity of the climate and the inaccessibility of these regions will preclude any large-scale settlement there in the foreseeable future.

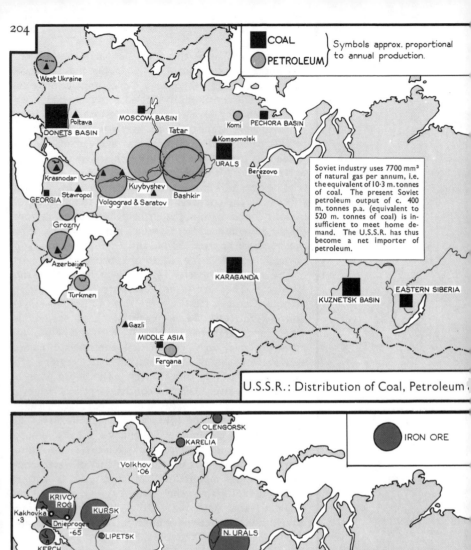

COAL

PETROLEUM

} Symbols approx. proportional to annual production.

West Ukraine

DONETS BASIN
Poltava

MOSCOW BASIN

Komi

PECHORA BASIN

Tatar

Komsomolsk

URALS

Berezovo

Krasnodar

Stavropol
GEORGIA

Volgograd & Saratov

Kuybyshev

Bashkir

Grozny

Azerbaijan

Türkmen

Gazli
MIDDLE ASIA

Fergana

KARAGANDA

KUZNETSK BASIN

EASTERN SIBERIA

Soviet industry uses 7700 mm³ of natural gas per annum, i.e. the equivalent of 10·3 m. tonnes of coal. The present Soviet petroleum output of c. 400 m. tonnes p.a. (equivalent to 520 m. tonnes of coal) is insufficient to meet home demand. The U.S.S.R. has thus become a net importer of petroleum.

U.S.S.R.: Distribution of Coal, Petroleum

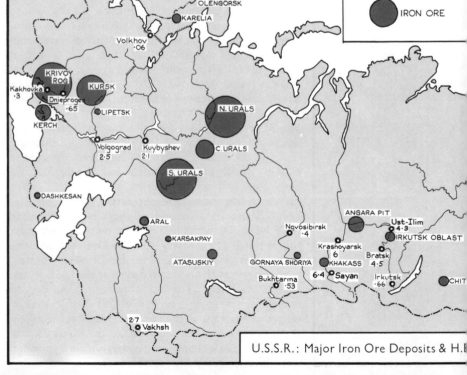

OLENGORSK

KARELIA

IRON ORE

Volkhov ·06

KRIVOY ROG

KURSK

Kakhovka ·3
Dnieproges ·65

LIPETSK

N. URALS

KERCH

Volgograd 2·5

Kuybyshev 2·1

C. URALS

S. URALS

DASHKESAN

ARAL

KARSAKPAY

ATASUSKIY

ANGARA PIT

Novosibirsk ·4

Ust-Ilim 4·3

IRKUTSK OBLAST

Krasnoyarsk 6

Bratsk 4·5

GORNAYA SHORIYA

KHAKASS

Bukhtarma ·53

6·4 Sayan

Irkutsk ·66

CHIT

2·7 Vakhsh

U.S.S.R.: Major Iron Ore Deposits & H.E.

These maps show the principal mineral and power resources of the Soviet Union. Notice the very large and widespread occurrence of such vital deposits as coal, iron ore, petroleum and natural gas. Vast untapped reserves of these minerals await development especially in Middle Asia and Siberia. In recent years there has been a great expansion in Soviet industry, total output having expanded, for example, by no less than 29% in the years 1966–1968. Such rapid progress has been made possible by spectacular achievements in all branches of the steel industry: (7) draw bar diagrams to illustrate the following output figures (*all million tonnes*):—

Year	Pig Iron	Steel	Sheet metal	Pipes
1921	0·1	0·2	0·1	Nil
1932	6·2	5·9	4·0	0·3
1940	14·9	18·3	11·4	0·96
1950	19·2	27·3	18·0	2·0
1960	46·7	65·3	43·7	5·8
1969	83·4	112·6	78·0	11·5
1970 (estimated)	96·0	127·0	97·0	15·0

Future emphasis is to be placed on the production of special steels and the modernization of equipment, especially in the rolling mills. In the key engineering industries the main aim will be to produce a great range of sophisticated machinery such as that listed below.

These high-quality machines will make it possible to reconstruct the entire Soviet economy, and special attention will be paid to streamlining and expanding the power, chemical and metallurgical industries. Two giant steel projects are already under construction, one in Kursk and another in the Yenesei Basin: both will use immensely rich local resources of iron ore and coking coal.

Modern trends in the Soviet steel and engineering industries include:—

the construction of automated blast furnaces, steel conversion mills, rolling shops and tin-plate works;

the manufacture of precision instruments computers, machine tools, new farm machinery and excavation gear.

The continuous rapid growth of industry in the Soviet Union means that more and more of its citizens live in towns. This is most noticeable in European Russia and in the Kuznetsk Basin where large numbers of agricultural workers, displaced from the land by farm mechanization, move into the towns to take jobs in factories. This trend is apparent even in traditional agricultural strongholds such as the Ukraine and the Central Black Earth Region (*see pp. 210–213*). On the other hand there have recently been significant increases in the *rural* population in Kazakhstan and Middle Asia as thousands of pioneer settlers have moved into these newly reclaimed wheat growing lands. (*See pp. 230–231*).

THE MOSCOW–GORKY REGION, centrally placed on the Russian part of the European Plain, was the focus of the old Russian Empire. From the 15th century onwards the Russians, a Slav people, gradually extended their power from this centre to become the strongest group in the territories which now form the U.S.S.R. This expansion involved them in centuries of warfare with a great many peoples including Tartars, Turks, Poles, Georgians, Armenians and Mongols, all of whom they eventually dominated. In 1917 the Russian Empire was torn by revolution and counter-revolution, but by 1923 the Communist-ruled 'Soviet Union' was firmly established in its place. Moscow, as the capital of Russia, naturally became the very heart of the new state; now it is the cultural and economic focus of the U.S.S.R. and the main city in a major industrial region.

The limits of this region are not clearly defined, but may be regarded as a circle with a radius of approximately 240 km centred south-west of Vladimir (*see map*). About 70% of the 21 million people within this region are town-dwellers, and more than half of them live within 96 km of Moscow. The only other very large town in the region is Gorky (1 125 000), but (*8*) how many towns are there with a population of over 50 000? This conglomeration of manufacturing towns makes the Moscow–Gorky region the most urbanised and industrialised portion of the U.S.S.R. Some 10% of the country's population live here and these towns produce about 25% of the total Soviet industrial output.

The region is not well suited to farming. In glacial times it lay beneath the Scandinavian ice-sheet (*map p. 11*) and the sticky boulder clays left behind when the ice melted now form extensive patches of flat, swampy land. Winters are long and cold, and the growing season relatively short. Flax, potatoes, hay and hardy grains such as oats and rye are grown throughout the region, and cattle are important on the damp, cool lowlands north of the Smolensk–Moscow Ridge.* Recently there has been a big increase in pig and poultry raising, the animals being fed on waste products from food-processing industries, e.g. potato-alcohol distilling. Market-gardening is important around all the larger towns.

There are iron ore deposits around the city of Tula, and lignite (brown coal) is mined on a large scale near Moscow and used as

*This ridge consists of a huge *terminal moraine*.

fuel in thermal-electric power stations, but otherwise the Moscow–Gorky region contains no important minerals. Thus the region's remarkable economic growth is not due to outstanding natural resources. It became a centre of relatively dense Russian settlement in the Middle Ages, mainly because its forests gave protection against the nomadic Tartar hordes of the southern grasslands. Also, in times when goods and people moved mostly by water transport, this region controlled the headwaters of the River Volga —the most important inland waterway in the country. Most towns in the region originated as fortified river-ports. (9a) On *which rivers are each of the following towns located: Kalinin, Moscow, Yaroslavl, Kostroma, Ryazan, Dzerzhinsk, Gorky?* (9b) Why do you think the position of Gorky especially favoured its growth as a commercial port?

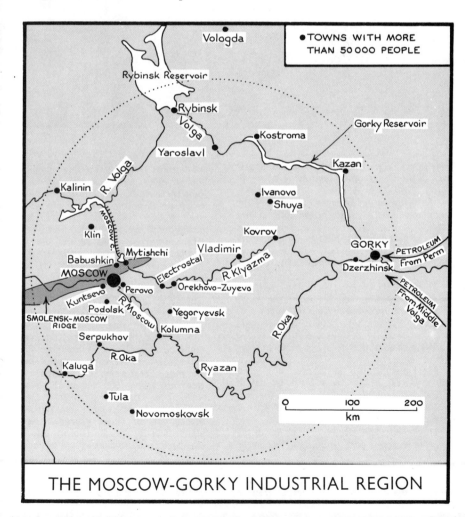

THE MOSCOW-GORKY INDUSTRIAL REGION

THIS IS MOSCOW . . .

—with over 7 million inhabitants the largest
city in the Soviet Union, and home of one in
thirty of all its people;

—seat of the Soviet Government, home of the
Premier and nerve centre of the entire Soviet
'bloc';

—the largest industrial producer and the main
cultural and university centre of the country;
an historic city and a major inland port.

This view shows the Moskva River, with the
wide embankment which carries the heavy
traffic that is banned from other central
streets. Notice that there is not a single
private car in sight—they are still rare, and
to Western eyes the streets of Moscow seem
almost empty.

In the distance is a 'skyscraper' of government offices in the heavy style typical of
Soviet architecture. On the right is the red-
walled, golden-domed Kremlin, centre of
government for one-sixth of the world.

The Kremlin (or 'citadel') which lies at the
heart of Moscow, was first built in 1147 on the
north bank of the Moskva River.

At first its high walls enclosed all the
government buildings, churches and dwellings of the city, but as the latter grew, a
series of new walls were built to protect the
inhabitants from marauding nomads. Although modern Moscow has spread far beyond these walls, their imprint remains in the
city's street pattern, for the old walls have
been pulled down and replaced by wide
boulevards. Beyond the outer boulevard
(the Sadovoye Koltso) lie the main railway
stations. Links to the city centre are provided by an underground railway network
renowned for its palatial and richly decorated
stations and escalators. The present city
limits are marked by a new ring-road, beyond
which lies a wooded residential 'green belt'
about six miles wide.

Moscow is a remarkable city, famous for its
many art galleries, museums, libraries,
cathedrals, palaces, government buildings
and magnificent parks and squares. Ancient
churches with gilded domes lie side-by-side
with huge modern department stores, and
Red Square, with Lenin's mausoleum, is a
Mecca for all Soviet citizens. Increasing
numbers of foreign tourists are also visiting
the Soviet capital, bringing an additional
source of income to its inhabitants. The
great majority of Moscow's working population, however, are employed in textiles,
engineering, metal-working and chemicals.

The rivers provided power as well as transport, and as early as the 18th century the string of towns along the Volga had grown into a major manufacturing district. Similar developments took place along the valleys of the Rivers Oka, Moscow and Klyazma. Details of the region's main industries are given in the notes.

The Moscow–Gorky region is vitally dependent on transport, for it receives foodstuffs and industrial raw materials from all over the U.S.S.R., and sends out manufactured goods in return. Railways carry over 80% of this traffic, but inland waterways, notably the Volga, also play an important role. Write an account of the main inland waterways in this region from the details given on the maps on pages 207, 204 and 222.

Gorky, together with nearby industrial cities such as Dzerzhinsk, forms a conurbation of over 1 million, the third largest metropolitan centre in the U.S.S.R. The city (formerly known as Nizhniy Novgorod) is favourably situated at the junction of the Rivers Oka and Volga. In the days of the Russian Empire it grew up as an important river port, renowned for its annual trade fairs. Today Gorky is the Soviet Union's main automobile manufacturing centre and has a wide range of food-processing and engineering industries, e.g. ships, aircraft and machine-tools. The adjoining cities of Dzerzhinsk and Balakhan have large chemical and paper industries respectively. (*10*) From the map on page 207 state the two sources of petroleum brought to Gorky.

MAIN INDUSTRIES IN THE MOSCOW–GORKY REGION

Textile industries were the first to develop, becoming particularly important in Moscow and in the towns along the Volga. Ivanovo, with its concentration on cheap cotton textiles, has long been known as the 'Manchester of Russia'. Formerly much raw cotton was imported, but in recent years the industry has been able to rely almost entirely on home-produced cotton from Soviet Middle Asia and Caucasia (*pp. 226 and 229*). The Moscow–Gorky region still produces about 80% of Soviet cotton textiles despite the development of industry in the cotton-growing areas. There is also an age-old linen industry based on locally-grown flax. The main centre of linen production is Kostroma.

Engineering was fostered in the 19th century by the development of iron and steel industry around Tula, but today large quantities of iron and steel are brought into the region from the Urals and Donbas steel-making centres (*pp. 213 and 221*). The main engineering centre is Moscow; other cities with specialised engineering production include Gorky (ships, automobiles), Kolomna (railway engines) and Kalinin (railway rolling stock).

Food-processing is important in nearly all the cities in this region. Such industries include fruit and vegetable canning, flour-milling, potato-alcohol distilling and sugar refining (from sugar-beet).

Chemicals have been manufactured in this region for many years, using local supplies of potatoes and grain to produce ethyl alcohol, and local phosphate rock to make fertilisers. Recently a large petro-chemical industry has developed in Dzerzhinsk using petroleum and natural gas piped from the Caucasus, the Ukraine and the Volga fields.

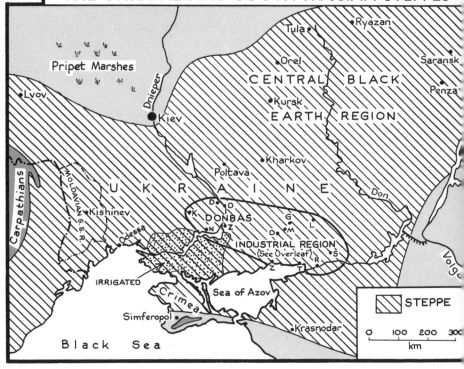

THE UKRAINE and SOUTH RUSSIAN STEPPES. Look again on page 200 at the photograph showing part of the great European Plain. The southern portion of this immense lowland includes the Steppes—flat or gently rolling country stretching from the Carpathians to the Volga. (*For details of climate, soils and natural vegetation, see p. 23.*) The Steppes cover most of the Ukraine, and in the north-east continue into the famous Central Black Earth country, which contains some of the finest farmland in the Soviet Union. Although this region often suffers from drought, especially in the south and east, it has enormous agricultural resources. It is also rich in minerals and includes a major heavy industrial zone—the Donbas. Some idea of the importance of the Ukraine in the Soviet economy can be gained from the facts shown on the left.

Agriculture on the Steppes is dominated by arable farming, notably the growing of cereals. Mile after mile is under wheat or maize, and large-scale

% OF CERTAIN SOVIET COMMODITIES PRODUCED IN THE UKRAINE			
Maize	57	Coal	33
Sugar-beet	55	Coking-coal	56
Sunflowers	42	Iron Ore	50
Wheat	10	Manganese	50
Meat	20	Steel	40
Butter	25	Pig Iron	50

The Ukraine contains only 3% of the area of the U.S.S.R., but 19% of its population. 53% of Ukrainians are town dwellers.

Harvesting wheat in the Ukraine.
(See also photo p. 23.)

mechanised cultivation is the rule. The landscape closely resembles that of the Prairie Provinces of Canada, and some of the State Farms, employing hundreds of work-people, are as big as English counties. There are also many 'collective farms' on which peasants own small plots around their homesteads, but are mainly concerned with working huge communally-owned estates. (*II:118–127 and III:148*.) Other details are given below.

(*11*) Copy a *large* outline of the map shown opposite. (*12*) Draw a line across your map through Kishinev and Kharkov; this line coincides approximately with the 500 mm isohyet. On your map (*13*) put the labels N.W. HUMID STEPPES and S.E. DRY STEPPES, to the north and south respectively of the 500 mm isohyet. Then, for each of these regions (*14*) add the main facts given in the notes below.

NORTH-WESTERN HUMID STEPPES	SOUTH-EASTERN DRY STEPPES
RAINFALL EXCEEDS 500 mm (625 mm at Lvov).	Except in Crimean Mts., RAINFALL BELOW 500 mm. (200–400 mm along Black Sea coast.)
SUMMERS COOLER and CLOUDIER than in south. WINTERS VERY COLD.	SUMMERS HOT, with clear skies and frequent scorching S.E. winds. WINTERS COLD.
Kiev: Jan. −6°C., July 19°C., Rainfall 525 mm. Former vegetation: wooded steppe or broad-leaf forest.	Odessa: Jan. −3°C., July 23°C., Rainfall 400 mm. Former vegetation: steppe grassland, becoming scrubby towards south.
Soils:—HUMUS-RICH BLACK EARTHS (chernozems), + FERTILE LOESS.	Soils: BLACK EARTHS + LOESS, deteriorating towards S. and S.E.
Main crops include: SUGAR-BEET FLAX POTATOES VEGETABLES CEREALS (Notably MAIZE and *SPRING* WHEAT).	Main crops include: CEREALS (Notably MAIZE and *WINTER* WHEAT); LINSEED; SUNFLOWER.
Many CATTLE and PIGS, esp. in N.W.	Many SHEEP and GOATS.

Throughout the Steppes dairying, poultry-farming and market-gardening are important around the larger towns. The chief drawback to agriculture is the cool and droughty climate. At present the only important irrigation project in the region lies around the Southern Dnieper, using water from the Kakhovka Reservoir. (15) Show on your map the irrigated area indicated opposite.

Right: *Iron ore being loaded on to railway trucks at Krivoi Rog.*

The Ukraine produces 8% of the world's output of steel and 10% of its output of pig iron. The annual steel output exceeds that of Great Britain or West Germany.

Hydraulic mining of coal in the Novo-Gorlovka Mine in Selidovo, the Donbas.

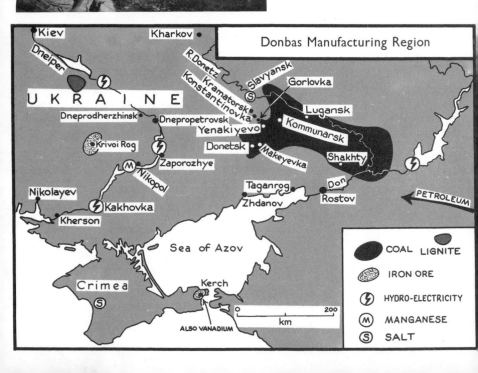

Donbas Manufacturing Region

Kiev
Kharkov
Dneiper
R.Donetz
Slavyansk
Kramatorsk
Gorlovka
Konstantinovka
U K R A I N E
Lugansk
Dneprodherzhinsk
Dnepropetrovsk
Kommunarsk
Yenakiyevo
Kommunarsk
Krivoi Rog
Donetsk
Makeyevka
Shakhty
Zaporozhye
Nikopol
Nikolayev
Taganrog
Don
Kakhovka
Zhdanov
Rostov
PETROLEUM
Kherson
Sea of Azov
Crimea
Kerch
ALSO VANADIUM

COAL LIGNITE
IRON ORE
HYDRO-ELECTRICITY
MANGANESE
SALT

0   200
km

The Industrial Ukraine. The photographs here and overleaf illustrate the mineral and power resources which have enabled the Ukraine to become the chief industrial region of the U.S.S.R. Mining began on the Donetz coalfield in the 1840's, and by 1913 the 'Donbas' was producing 87% of the country's coal. Today, despite competition from newer and richer coalfields, the Donbas still leads in output, producing 33% of the coal and 57% of the coking-coal of the U.S.S.R.

Mining in the Donbas is difficult because the more accessible seams have been worked out and those that remain are rather thin and lie deep below ground (nearly 1000 m down in places). However, the high quality of the coal makes mining worth while, several important varieties being produced. (*16*) Add labels to an outline map of the Donbas coalfield to show that the following types of coal are mined:—*anthracite* in the east; *coking coal* in the centre and west; *gas* and *long-flame coals* in the extreme west and south-west.

The Ukraine also has major deposits of other minerals, the chief of which are indicated on the map. (*17*) Name and locate each deposit. In the 19th century an iron and steel industry developed in the Donbas, using 'black-band' iron ore from the coalfield. Then in 1884 a railway link was completed between the coalfield and the vast high-grade ironfield at Krivoi Rog. This field soon became the main supplier of iron ore to the Donbas, but ores are also brought by ship and train from Kerch. Return cargoes of coking-coal are sent from the Donbas to Krivoi Rog and Kerch. The result of this 'shuttle-service', as the table shows, is that iron and steel mills have grown up in three types of town. (*18*) Show this on a map by underlining the towns in three separate colours, according to their categories. Add a key.

MAIN IRON AND STEEL TOWNS OF THE UKRAINE

Coalfield Towns		Ironfield Towns	Towns on routes between Coalfields and Ironfields	
Donetsk	Makeyevka	Krivoi Rog	Dneprodherzhinsk	Nikopol
Yenakiyevo	Konstantinovka	Kerch	Taganrog	Zaporozhye
Kramatorsk	Kommunarsk		Zhdanov	Dnepropetrovsk

The engineering and chemical industries are also very important in the Ukraine. *Engineering* industries are established in the steel-making centres (i.e. in all the larger Donbas towns), and also at many regional centres throughout the Ukraine. The main engineering towns are shown overleaf, together with the goods

Kramatorsk (*equipment for the iron and steel industry*)	Nikolayev (*shipbuilding*)
	Rostov-on-Don (*agricultural machinery*)
Kharkov (*transport, electrical and agricultural engineering*)	Lugansk (*steam and diesel locomotives.*)
	Kherson (*shipbuilding*)
Gorlovka (*coalmining equipment*)	Kiev (*agricultural and general engineering*)

they produce. (*19*) Put this information on a *large* sketch-map, ringing the Donbas towns.

Chemical manufacturing towns in the Ukraine fall into three main categories, as follows:—(i) *coalfield towns*, e.g. Makeyevka, Donetsk and Gorlovka, where coal-tars and gases derived from coke are used as raw materials; (ii) *towns with large h.e.p. plants* where electro-smelting industries have been established, e.g. aluminium smelting at Zaporozhye; (iii) *towns on the salt-field*, e.g. Slavyansk, where large amounts of sodium salts are mined and acids, fertilizers, synthetic fibres and other chemicals are made. (*20*) Add these chemical industrial centres to your map of engineering towns, underlining them in an appropriate colour. Give your completed map a key and a title.

In recent years the industrial Ukraine has used more and more h.e.p. to supplement electricity supplies from coal-fired power stations. This view shows part of the great dam and power station on the Dneiper at Kakhovka. Three similar dams exist farther upstream and two more are planned. ((*21*) *Where?*) Although these dams are intended primarily to improve irrigation and navigation, their power stations are among the largest in the country. Large quantities of h.e.p. are used in the industrial towns of Dneprodhzerzhinsk, Dnepropetrovsk and Zaporozhye.

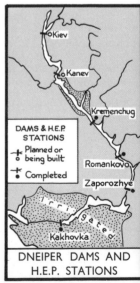

DAMS & H.E.P. STATIONS
+ Planned or being built
+ Completed

DNEIPER DAMS AND H.E.P. STATIONS

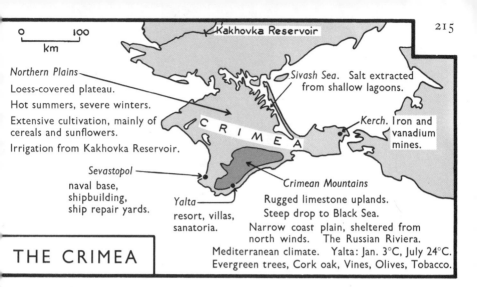

THE CRIMEA

Kakhovka Reservoir

Northern Plains
Loess-covered plateau.
Hot summers, severe winters.
Extensive cultivation, mainly of cereals and sunflowers.
Irrigation from Kakhovka Reservoir.

Sivash Sea. Salt extracted from shallow lagoons.

Kerch. Iron and vanadium mines.

C R I M E A

Sevastopol
naval base,
shipbuilding,
ship repair yards.

Yalta
resort, villas,
sanatoria.

Crimean Mountains
Rugged limestone uplands.
Steep drop to Black Sea.
Narrow coast plain, sheltered from north winds. The Russian Riviera.
Mediterranean climate. Yalta: Jan. 3°C, July 24°C.
Evergreen trees, Cork oak, Vines, Olives, Tobacco.

0 100
km

The Crimea is a diamond-shaped peninsula, forming part of the Ukraine but sufficiently distinctive to be mentioned separately. The main details of the geography of the Crimea are given on the map above. (22) Use that information to write a short geographical account of the region.

THE MOLDAVIAN REPUBLIC (*map p. 210*) was formed in 1940, mainly from land annexed from Rumania. It consists very largely of steppe and is the most densely populated Republic of the Soviet Union. There are no large cities, however, for most of the people live in large villages and work on the land. The steppe soils, enriched with loess, are very fertile and the climate is comparatively mild. Grain crops, e.g. maize, winter wheat and barley, together with nuts, fruit and vegetables, are grown, but the most famous Moldavian crop is the vine. The country around Kishinev is virtually one huge vineyard (*see photo*), and this district has been called 'The Russian Champagne'.

THE VOLGA LANDS. This photograph shows a huge barge 'train' on the River Volga. Traffic on this 3680 kilometre-long river is increasing, for it is the major navigational waterway of the Soviet Union, comparable in importance to the Rhine in Western Germany. Note (*below*) that the territory crossed by the Volga varies from cool forested land in the north-west to semi-desert bordering the Caspian. In spite of differences in climate, soils, vegetation and land-use, however, the riverine lands are unified by the Volga to such an extent that they are known simply as the *Povolzhye*—'along the Volga'. Trade is so important along the Volga because of the course of the river in relation to three flourishing industrial centres, i.e. the Moscow region, the Donbas and the Urals. This is indicated in the diagram below. (*23*) Make a *large* copy of the sketch-map, adding the following information by labelling the regions of origin of the cargoes:—

THE VOLGA BASIN (Povolzhye)

Cool Humid

Mixed Forest Land

150m

• Kirov

Perm Ⓟ

Ⓗ Gorky

Volga

Kazan •

Ⓟ

Foothills of URALS

Ⓟ Almetyevsk

Ufa Ⓟ

Volga Hts.

Deciduous Forest & Arable

Ph Ⓗ

Syzran • Ⓢᵤ

Ⓟ Kuybyshev

Steppe

Ⓖ

Saratov

Ⓟ Kamyshin

Don

Ⓖ

Ⓟ Volgograd

Volga

Ⓢ

Flat Desert Plain (Former Floor of Caspian)

Astrakhan •

0   200
km

Caspian Sea

Ⓗ H.E.P.

◯ LIMIT OF OILFIELD

Ⓟ PETROLEUM

Ⓖ NATURAL GAS

Ⓢ SALT

Ⓢᵤ SULPHUR

Ph PHOSPHATE

THE VOLGA WATERWAY

◯ INDUSTRIAL REGION

Sverdlovsk

• Perm

URALS

Volga

Kama

Magnitogorsk

Gorky

• MOSCOW

Kuybyshev

Tula

Volga

Don

Volgograd

Volga

DONBAS

Rostov

Caspian Sea

Baku

Black Sea

0   300
km

Cargoes travelling *UPSTREAM* include:—*wheat, coal and pig-iron from the Ukraine; fish from the Caspian; salt from the lower Volga; refined petroleum products from the Caucasus.*

Cargoes travelling *DOWNSTREAM* include:—*timber from the northern forests; manufactured goods from the Moscow region; mineral ores from the Urals; crude petroleum from the Middle Volga oilfield.*

Note the place of the new Volga–Don Canal in this trade pattern. Navigation on the Volga has been enormously improved in recent years by the construction of a whole series of gigantic dams (*see map*). These dams, like those on the River Don, provide water for irrigation and also make possible the production of huge quantities of hydro-electricity. So much electricity is now generated in the Povolzhye that this region has been called 'the greatest power-house in the Soviet Union'.

The towns which grew up on the banks of the Volga were originally river ports and trading centres. Today many of them have grown into great industrial cities, notably those at railway bridging points, e.g. Volgograd, Saratov, Kuybyshev and Kazan.

"Since 1940, the Middle Volga and adjacent 'pre-Ural' region has been the most rapidly developing region in the country in terms of industrial production, trade, and major cities. . . . Railroad freight also has increased the most rapidly in the Middle Volga pre-Ural area."[*]

Modern industrial development of the Volga lands was triggered off during the Second World War. As their western industrial cities were overrun by the Germans the Soviet people made desperate efforts to move their equipment and factories eastwards. At one time, when Moscow was under siege, even the Central Government was moved to Kuybyshev. Large-scale industrialisation was made possible in the Middle Volga region by the discovery of immensely rich oilfields between Volgograd and Perm (*see map*). Since 1956 these fields have produced over 65% of the country's petroleum. Kuybyshev and Syzran have become major oil-refining centres: refined products are sent by pipeline as far

[*] Paul E. Lydolph, *Geography of the U.S.S.R.*

Kuybyshev (998 000) is the largest city on the Volga after Gorky, and the focal centre of the entire Volga region. The city is sited at the point where the small Samara River enters the Volga from the southeast. In pre-Soviet times Kuybyshev (formerly called Samara) was the financial capital of the Volga wheat trade. Today it Is a vital bridging point of the Volga and a fast-growing industrial city, with major engineering, oil-refining, chemical and food-processing works.

Kazan (824 000) is the second largest city in the Povolzhye. Its rapidly developing industries are primarily concerned with chemicals and engineering, notably the manufacture of office equipment such as calculating machines and typewriters. At one time Kazan was an important market for cattle, hides and furs, and it has old-established leather and shoe-making industries.

Volgograd (750 000) was until recently called Stalingrad, and is one of the most famous cities in the Soviet Union. It was in Stalingrad that the German onslaught was held in a ferocious battle that proved to be the turning point of World War Two. Originally built as an outlying Russian fortress, the city became a thriving port and then a metal-working centre using iron and steel from the Donbas. Today Volgograd has very large metallurgical, chemical, lumber, food-processing and engineering industries (the city is renowned for its production of tractors). New industries include electro-chemicals and aluminium smelting, using h.e.p. from the Volga. The Volga–Don Canal, opened in 1952, has greatly increased waterborne trade through Volgograd.

Saratov (725 000) is another major railway crossing-point on the Volga, carrying a main line from Moscow to the Urals industrial region. The city lies at the centre of a large natural gas field. Its industries include engineering, chemicals and food-processing.

Astrakhan (371 000) is primarily a port for the Northern Caspian, handling local products such as fish and salt and petroleum products from the Caspian oilfields (p. 227). Astrakhan lies off the main trade route linking the Donbas to the Moscow region via the Middle Volga, and although it has engineering and food-processing industries it has not developed on such a large scale as the cities farther upstream.

Assembly line. Volgograd tracto
factory.

afield as Leningrad and Irkutsk. Natural gas from an important field near Saratov is also piped to Moscow. A large petro-chemical industry has developed in the Volga region, based on natural gas and by-products from oil refining: the main products are synthetic rubber, fertilisers, artificial fibres and alcohol.

Other resources of the Povolzhye are shown on the map page 216: chemical and fertiliser industries use the phosphoric rock and sulphur deposits near Kuybyshev, and salt from Lake Baskunchak is used in the important sturgeon-fishing industry of the Northern Caspian. Details of the main industrial cities along the Volga (other than Gorky) are given opposite. (*24*) Draw a sketch-map to show their position and add short notes about each one. (For Gorky, see page 209.)

Despite the large-scale and rapid industrialisation taking place at various river ports, much of the Volga region is still pre-dominantly rural. Agriculture is handicapped, however, for in the north the soils are infertile and badly drained and the growing season is short, while the south degenerates into desert. In the Middle Volga region between Kazan and Volgograd there are distinct contrasts between the climate and the appearance of the land on opposite banks of the river. Details of these contrasts and the differences in land-use which result are shown in the dia-gram below. (*25*) Use the information given there to write a short account of farming in the Middle Volga lands. The lower Volga flood-plain between Volgograd and Astrakhan is the scene of intensive irrigation farming, where rice, fruit, melons and some cotton are grown. Outside this green, irrigated strip the land is useful only for dry grazing.

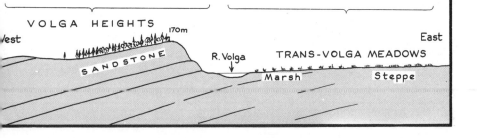

Rugged sandstone uplands

Rainfall > 350 mm. Relatively cool (Temperatures modified by altitude). Forested: conifers and mixed decidu-ous trees.
Mixed farming between Kazan and Saratov. Wheat, hay, maize, sugar-beet, some dairying.

Alluvial flood-plains

Very flat marshy lowland—the Trans-Volga Meadows—changing south-east-wards into steppe and semi-desert. Rainfall < 350 mm. Western edge of the ' Virgin Lands ' (see p. 230). Much spring-wheat grown here since 1954.

VOLGA HEIGHTS
170m
West
East
SANDSTONE
R. Volga
TRANS-VOLGA MEADOWS
Marsh
Steppe

Machining a forging press at the Urals Heavy Engineering Works.

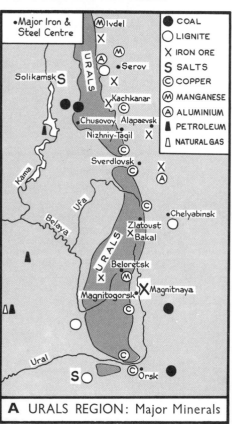

A URALS REGION: Major Minerals

- •Major Iron & Steel Centre

- ● COAL
- ○ LIGNITE
- X IRON ORE
- S SALTS
- © COPPER
- Ⓜ MANGANESE
- Ⓐ ALUMINIUM
- ▲ PETROLEUM
- ◺ NATURAL GAS

Ivdel
Serov
Solikamsk
Kachkanar
Chusovoy Alapaevsk
Nizhniy-Tagil
Sverdlovsk
Kama
Ufa
Belaya
Chelyabinsk
Zlatoust
Bakal
Beloretsk
Magnitogorsk Magnitnaya
Ural
Orsk
URALS

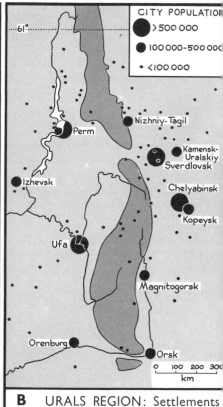

B URALS REGION: Settlements

CITY POPULATION
- ● >500 000
- ● 100 000–500 000
- • <100 000

61°
Perm
Nizhniy-Tagil
Kamensk-Uralskiy
Sverdlovsk
Izhevsk
Chelyabinsk
Kopeysk
Ufa
Magnitogorsk
Orenburg
Orsk

0 100 200 300
km

THE URALS form the traditional boundary between Europe and Asia. They extend southwards for more than 1900 kilometres from the Arctic Ocean, and represent the stumps of heavily eroded folded ranges. In the north the Urals consist of a single barren, glaciated ridge, which in places is more than 2000 m high. South of latitude 61°N., however, they split into a series of low, ill-defined, forested ranges, rarely exceeding 1000 m and easily crossed by roads and railways.

For centuries the Urals were regarded as marking the extreme edge of civilised Russia, but since the Revolution of 1917 they have been the scene of spectacular industrial developments based on rich and varied mineral deposits. The main deposits are shown on Map **A** (*opposite*): their relative importance is apparent from the maps on pages 204–5. (*26*) To what extent does the distribution of population in the Urals reflect the distribution of minerals? The principal mineral mined in the Urals is iron ore, notably in the famous *Magnitnaya* (Iron Mountain) near Magnitogorsk. Other high-grade iron ores are mined at Nizhniy-Tagil and Bakal, and there are immense low-grade deposits around the eastern and southern flanks of the Urals, e.g. at Kachkanar. Coal for fuel and steel-making is brought to the industrial cities of the Urals from the Kuznetsk Basin, over 1600 kilometres to the east, and from Karaganda, 960 kilometres to the south-east (*map p. 231*). Recently, however, very large deposits of petroleum and some natural gas have been discovered around Ufa, and there is natural gas in the Ob Basin near Berezovo. Pipelines also bring in petroleum and natural gas from Uzbekistan, so that one-half of the fuel needs of the Urals industries are now being met by oil and gas.

As in the Middle Volga region, industrialisation of the Urals was given a tremendous impetus during World War Two. Old-established metallurgical centres such as Zlatoust and Chusovoy were reconstructed and modernised, and huge steelworks and engineering factories have grown up at Magnitogorsk, Sverdlovsk, Nizhniy Tagil, Orsk and Izhevsk. The steel towns of the Urals now produce about one-third of the Soviet Union's pig-iron and steel. Although metal-using industries are expanding, over 65% of the rolled steel produced in the Urals is at present moved out for use elsewhere, notably in the Moscow and Middle Volga regions. There are chemical industries in Perm, Sverdlovsk and Solikamsk which use local supplies of sulphur and salts.

(27) On a *large* copy of the map

(a) Shade yellow the land north of the line T–T′, and label it DESOLATE TUNDRA.

(b) Shade dark green the land between the line T–T′ and 60° N. and label it DENSE CONIFEROUS FORESTS: MUCH LUMBERING.

(c) Shade the remainder of the land area light green, and label it UNDULATING GLACIAL PLAIN: MIXED FARMING—OATS, FLAX, HAY, POTATOES, DAIRYING and PIG-REARING. 25% FORESTED.

(d) Indicate that the ridge running eastwards from Smolensk is a huge TERMINAL MORAINE and that it carries the main road and railway from Warsaw to Moscow.

(e) Indicate that the Barents and White Seas are important FISHING GROUNDS and that the towns on the White Sea coast are the main fishing ports.

(f) Show that there are large, worked deposits of:—PHOSPHATES at Kirovsk; ALUMINIUM ORES at Kirovsk and Tikhvin; NICKEL and COPPER at Pechenga; POTASH at Starobin.

(g) Label the Baltic–White Sea Canal and name the large lake through which it passes (use an atlas).

(h) Label the marshy area south of Smolensk PRIPET MARSHES, and suggest the probable cause of this marshy ground.

(i) Indicate that the Gulf of Finland is closed by ice from December until May.

The River Neva at Leningrad. The buildings—formerly palaces of the Tzarist nobility—are now museums.

THE EUROPEAN NORTH and WEST is a large region, but economically it remains one of the least developed parts of the U.S.S.R. The reasons are largely geographical, the main disadvantages being (i) the climate, which varies from cool and moist in the south to bitterly cold tundra in the north and (ii) the glaciated lowland terrain; nearly all of the area shown on the map was scoured by the great Scandinavian ice-sheets, which scraped away the soils and left behind an immense load of morainic boulders, sands and gravels. Farming and transport are hindered by the poor, infertile soils and rocky outcrops, as well as by extensive swamps and thousands of lakes on undrained boulder clay.

Even so the region has great timber and mineral resources and there are rich fishing grounds in the White and Barents Seas. Moreover the Arctic ports provide the Soviet Union with its only direct ice-free Arctic sea routes to North-West Europe and the Atlantic. Much of the area shown on the map is sparsely settled, especially in the north, but there are several important cities.

Leningrad (3 715 000) is the second largest city in the Soviet Union. It was founded in 1713 by Peter the Great, who chose its position deliberately to force Russia into closer contact with the life and ideas of Western Europe. In spite of its island site in the swampy delta of the Neva River, surrounded by a forested wilderness, the city prospered and from 1713 until 1919 it was the capital of the Russian Empire. Formerly called St. Petersburg (and for a while Petrograd) it was re-named Leningrad in 1924, to commemorate the principal leader of the Soviet Revolution. The city has many famous monumental buildings, an inheritance from Tsarist times, which together with fine modern boulevards and parks make it a showpiece for tourists.

In the 1880's St. Petersburg became the first Russian city to feel the full impact of the Industrial Revolution ((*28*) *suggest why*) and today it has big engineering, metallurgical, chemical, textile and food-processing industries. About 10% of the steel output of the Soviet Union is used in Leningrad's factories. (*29*) Draw a sketch-map to show that the city obtains: *coal* from Vorkuta, *iron ore* from Karelia and the Kola Peninsula, *refined oils and natural gas* from the Caucasus, *cotton* from Middle Asia and many *foodstuffs* from the Ukraine. The city has a large steelworks and a shipbuilding industry and is a major Soviet naval base.

The need for Leningrad to import industrial raw materials and foodstuffs from all over the Soviet Union makes the city very vulnerable in wartime. In World War Two it withstood a terrible siege for nearly three years, a 'Road of Life' being kept open in winter across the frozen surface of Lake Ladoga. As a result of their wartime experiences the Soviet Government prohibited migration into very large cities and it is therefore unlikely that Leningrad will undergo further expansion.

Archangel (315 000) was established in 1583 as a port for trade between England and Muscovy (as the Moscow region was called in those days). The city has always handled tundra and forest products: in the past these were mostly skins and furs, but today Archangel is the most important lumber port in the Soviet Union. The coniferous forests of the European North yield 13% of the country's timber, mostly spruce and Scotch pine. Throughout the summer logs are floated downstream to Archangel along the many headstreams and tributaries of the Northern Dvina (*photo opposite*). This activity is on such a large scale that the Dvina takes second place amongst all Soviet rivers in respect of freight moved. ((*30*) *Which river ranks first?*) In Archangel the logs are handled by more than 150 sawmills and the timber products. are shipped out via the White Sea in the following spring.

Murmansk (290 000) is the largest town north of the Arctic Circle. It owes its importance (*a*) to its ice-free position (*re-read pp. 14–18 and (31) explain why Murmansk is accessible to ships throughout the year whereas Leningrad, 9° farther south, is ice-bound from December until May*), and (*b*) to the growth of a large fishing fleet operating in the Barents Sea. Murmansk is linked to Moscow by rail and replaces Leningrad as the northern commercial port of the U.S.S.R. during the mid-winter freeze-up. In addition Murmansk exports phosphates from the mines at Kirovsk, and is an important naval base. It has shipbuilding and ship-repair yards (*II:104-8*).

MAIN INDUSTRIES IN THE BALTIC REPUBLICS AND BELORUSSIA

Food-processing, e.g. flour-milling, meat packing, sugar-beet refining and distilling.

Light industries, e.g. light-engineering, leather, furniture and textiles.

Saw-milling and the *production of wood-pulp*.

Fishing, both sea- and fresh-water.

THE BALTIC REPUBLICS of Estonia, Latvia and Lithuania once formed part of the Russian Empire. In 1919 they achieved a short-lived independence, but were absorbed into the U.S.S.R. in 1939, together with the Karelian Isthmus and parts of eastern Poland. (*Map p. 245.*) The Baltic and adjoining Belorussian territories (*map p. 222*) suffered enormous damage and casualties during the Second World War—hundreds of thousands of people were killed and large numbers of 'suspect' German-speaking inhabitants were deported to Siberia. In recent years, however, these population losses have partly been made good by a return flow of Russian immigrants.

Although the climate is relatively mild and moist in the Baltic and Belorussian lands, agriculture is hampered by infertile glacial soils and widespread bogs.

Details of farming activities are given in the notes on page 216. About one-quarter of the land is still forested.

Lack of raw materials and fuels have hindered the development of industry in these regions. Under Soviet rule, however, industrialisation has begun as indicated in the table. In addition there is some oil-shale between Tallinn and Narva; this yields natural gas which is sent by pipeline to Leningrad. The more important cities along the Baltic are *Tallinn* (saw- and pulp-mills, timber exports), and the seaport of *Riga* (timber and flax exports, textiles, shipbuilding, food-processing and engineering (e.g. machine tools and electrical equipment)). The main city of Belorussia is *Minsk* (775 000), which has developed important engineering industries using metals and fuels from other regions. Products include vehicles, ball-bearings, machine tools, radios and sewing machines.

Logs being floated downstream on the Northern Dvina.

Western Foothills ('*The Kuban*'). Subject to drought, but mean annual rainfall (625–775 mm) normally enough for cultivating a variety of crops without irrigation. Cold winters, hot moist summers. (Jan., *c.* −7°C.; July, *c.* 20°C.) WHEAT, SUNFLOWER, MAIZE, SUGAR-BEET. RICE grown in swampy Kuban R. delta. Some DAIRYING and LIVESTOCK breeding. Frequent desiccating winds from the Caspian deserts are a hazard to crops in summer.

Rioni Basin (*Colchis Lowland*). Low, swampy and very flat. Poor drainage. Mild winters, hot summers; rain in all seasons. Very wet in west (Batumi, 2250 mm). Many sub-tropical crops, notably CITRUS FRUIT, TEA, GRAPES, TOBACCO, MAIZE, MULBERRIES and RICE.

Eastern Foothills. Mostly semi-arid steppe and desert, used as dry grazing for sheep and camels. Irrigation vital for cultivation. Some WINTER WHEAT and MAIZE, RICE in Terek R. delta.

Araks Valley—irrigated COTTON.

Kura Lowland—steppe climate and natural vegetation. Irrigation essential for cultivation. COTTON, ALFALFA, ORCHARD FRUIT, GRAPES, MAIZE, MULBERRIES, RICE.

Caucasus Mountains and Armenian Plateau. Summer pastures for SHEEP and CATTLE. Animals go down to Rioni–Kura Lowlands in winter.

THE CAUCASUS region, bounded on the north by the Manych depression and on the south by the Turkish and Iranian frontiers, was brought under Russian control at the turn of the 19th century. Rostov-on-Don had been founded as a frontier outpost in 1761, and by 1828 the Russian armies had crossed the Caucasus Mountains and occupied Georgia and Armenia. To the south of the Greater Caucasus the population still consists mainly of Georgians, Armenians and Azerbaydzhanians, together with other small nationality groups. Russians predominate in the foothills to the north of the mountains.

Nearly 18 m. people live in the Caucasus region, most of them in the foothills or in the basins of the three main rivers—Rioni, Kura and Arak. The majority earn their living from agriculture; details are given in the notes above. (*32*) Use these notes to make a *large*, labelled sketch-map showing the main farming activities. The various sub-regions referred to are shown on the map below.

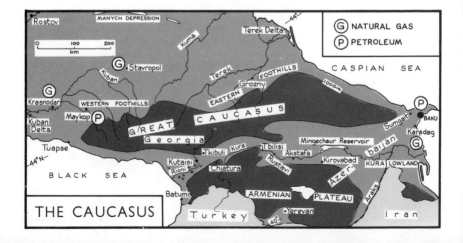

Pietr Polyansky is one of the 'New Georgians'. Though he comes from a family who for generations have worked on the land, Pietr went to a technical college in Tbilisi and studied engineering. Now he is employed in the 'New Town' of Rustavi (*see map*) in a steelworks making pipes and drilling equipment for the famous oil wells at Baku. For nearly a century Baku was the leading Soviet oilfield and, apart from local textile fibres, provided virtually the only basis for Caucasian industry.

(*33*) On a *large* copy of the map opposite add the following information by means of labels and symbols:—

(*a*) Oil pipelines run (i) *from Baku to Batumi via Tbilisi and Tkibuli*; (ii) *from Grozny to Rostov via Maykop*; (iii) *from Maykop to Tuapse*.

(*b*) Natural gas pipelines run (i) *from Karadag to Tbilisi via Kirovbad and Akstafa*; (ii) *from Akstafa to Yerevan*; (iii) *from Stavropol to Rostov*; (iv) *from Krasnodar to Rostov*; (v) *from Rostov to the Donbas and to Moscow and Leningrad*.

(*c*) Oil products from refineries in Batumi and Tuapse are shipped to the Donbas.

(*d*) Oil products from refineries in Baku are shipped to the Middle Volga.

(*e*) Oil products from refineries in Grozny and Maykop are sent by pipeline to Rostov and the Donbas.

Since 1945, with the development of the new Volga–Urals oilfield, the Caucasus fields have become less significant than formerly: today they produce about 15% of Soviet oil. Even so, oil and natural gas still play a very large part in the industrial life of the Caucasus region. In addition to the oil refineries mentioned above there are very large petro-chemical industries in Baku, Yerevan and Sumgait, and natural gas provides fuel for an increasing number of industrial towns. The latter include Tbilisi (textiles), Rustavi (iron and steel and engineering) and Kutaisi (automobile assembly). The Caucasus region also has the advantage of considerable h.e.p. (e.g. from the great dam at Mingechaur) and minerals such as copper, manganese and bauxite. Very large manganese deposits are worked at Chiatura, and Yerevan and Sumgait have rapidly expanding aluminium refining industries using h.e.p. and natural gas power respectively.

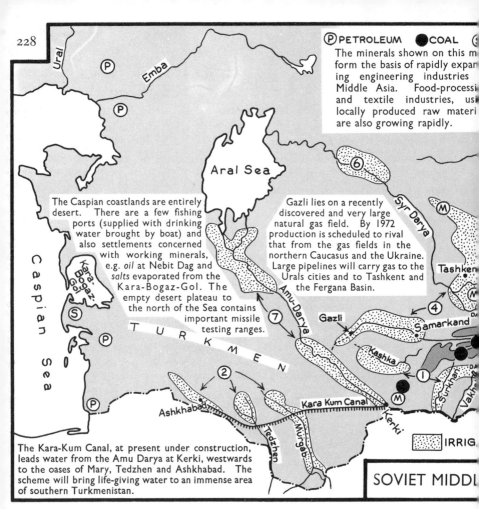

P PETROLEUM ● COAL

The minerals shown on this m
form the basis of rapidly expan
ing engineering industries
Middle Asia. Food-processi
and textile industries, usi
locally produced raw materi
are also growing rapidly.

Aral Sea

The Caspian coastlands are entirely
desert. There are a few fishing
ports (supplied with drinking
water brought by boat) and
also settlements concerned
with working minerals,
e.g. *oil* at Nebit Dag and
salts evaporated from the
Kara-Bogaz-Gol. The
empty desert plateau to
the north of the Sea contains
important missile
testing ranges.

Gazli lies on a recently
discovered and very large
natural gas field. By 1972
production is scheduled to rival
that from the gas fields in the
northern Caucasus and the Ukraine.
Large pipelines will carry gas to the
Urals cities and to Tashkent and
the Fergana Basin.

Caspian Sea

T U R K M E N

Gazli

Tashken

Kashka

Samarkand

Kara Kum Canal

Kerki

Ashkhabad

Tedzhen

Murgab

IRRIG

SOVIET MIDDL

The Kara-Kum Canal, at present under construction,
leads water from the Amu Darya at Kerki, westwards
to the oases of Mary, Tedzhen and Ashkhabad. The
scheme will bring life-giving water to an immense area
of southern Turkmenistan.

*SOVIET MIDDLE ASIA and SOUTHERN KAZAKH-
STAN* cover an area, almost as big as Western Europe. An atlas
shows that physically the region is very varied, consisting of plains,
basins, tablelands and mountains. Its 17 million inhabitants also
vary enormously, including primitive peoples such as the nomadic
Kirghiz shepherds, as well as sophisticated Russian immigrants:
the latter earn their living in the larger towns as engineers,
teachers, factory workers, and so on.

Despite this complexity of landscape and culture the region is
stamped by one common factor—lack of water: except for the
higher mountains it consists entirely of desert or steppe. Much
of it is uninhabited, being too dry, stony or mountainous to support life. Where there is water for irrigation, however, or where
rich mineral deposits exist, dense clusters of people are found.

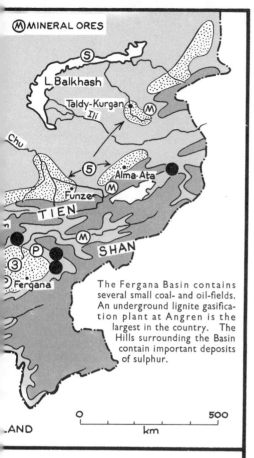

Ⓜ MINERAL ORES

L. Balkhash

Ⓢ

Taldy-Kurgan
Ili

Ⓜ

Chu

⑤

Ⓜ Alma-Ata

Funzer

TIEN

Ⓜ

Ⓟ

③ SHAN

Fergana

The Fergana Basin contains several small coal- and oil-fields. An underground lignite gasification plant at Angren is the largest in the country. The Hills surrounding the Basin contain important deposits of sulphur.

0   500
km

…AND

A & SOUTHERN KAZAKHSTAN

The main areas of irrigated farmland are indicated on the map and numbered 1–7. With the help of an atlas (*34*) identify each area from the following list:—*Fergana Basin, Tashkent-Samarkand, Tyan Shan Foreland, Amu-Darya, Syr Darya, Hindi Kush Foreland, Vasch-Surkhan-Kashka.* The very fertile loess soils and long, hot summers of these irrigated 'oases' favour intensive cultivation. Cotton (*III: 148–9*) is the principal crop of Middle Asia, but a great variety of other produce is grown including wheat, rice, fruit, vegetables, grapes and sugarbeet. New dams on the larger rivers (*see map*) supply increasing volumes of water for irrigation, and some of them also have power stations. Along the moister loess-covered foothills of the southern mountains dry farming for grain is practised (*III: 44*). Throughout Soviet Middle Asia efforts are being made to resettle thousands of nomadic shepherds on State Farms, where they are shown how to rear beef cattle and cultivate the land. Very large areas of poor mountain grassland remain as open sheep-grazing country.

(*35*) Use the information given on the map to write a short account of other economic activities in Middle Asia.

Below: *The Kara-Kum Canal (for details see map). Evaporation reduces the value of irrigation canals in such arid country as this.*

WESTERN SIBERIA and NORTHERN KAZAKHSTAN
form part of the region between the Urals and the Pacific which was
first brought under nominal Russian control during the early 17th
century. The first Russian immigrants were hardy fur-traders,
hunters and adventurers, but as time went by farmers settled in
the more favourable lands in the south, and by the late 19th century there was a belt of continuous rural settlement in Southern
Siberia and Northern Kazakhstan. In the 1890's the building of
the Trans-Siberian Railway gave a big impetus to settlement: new
bridging towns, which grew up at the main river crossings on the
Railway, soon developed food-processing, textile and lumbering
industries. More recently large mining and metallurgical centres
have grown up in the Kuznetsk Basin and in Northern Kazakhstan
—the principal mineral deposits involved are shown on the map.

In spite of these developments, however, very large parts of
Siberia and Kazakhstan remain desolate and uninhabited, and the
whole region—as the quotation opposite reveals—is still very much
one of pioneer settlement. Reasons why farming is at present
confined to a comparatively narrow 'corridor' in the southern part
of the region are indicated on the map: (*36*) write them down and
(*37*) estimate the approximate width of the belt of settled farmland. Although no part of the region is completely favourable for
cultivation, summer temperatures in the south are ideal for spring
wheat. Other noteworthy crops include barley, sunflowers, hemp
and sugar-beet, as well as hay and maize grown for silage.

Under the Virgin Lands Programme of recent years, wheat
cultivation has pushed steadily southwards into the arid grazing
lands of Kazakhstan. Tens of thousands of hectares of steppe-
grassland have been ploughed for the first time, the whole scheme
having been described as "wheat growing on the grand scale,
gambling on the weather".* The gamble, of course, lies in the
possibility of drought, for the rainfall is both small and unreliable.
Some critics of the Virgin Lands programme say that it is already
creating a huge 'Dust Bowl' similar to that which afflicted the
United States prairies in the 1930's, and they expect the Government will quietly let much of the ploughed-up land revert to grazing.

Efforts are being made to increase the area under irrigation, but
the only river in the main farming belt which does not dry up in
summer is the Irtysh. The main irrigation dams and reservoirs
are shown on the map. Hydro-electric plants and irrigation
reservoirs are also planned for the River Ob.

* Paul E. Lydolph, op. cit.

THE VIRGIN LANDS AWAIT YOUNG ENTHUSIASTS
Akmolinsk Province, Kazakhstan.

Dear Friends! Last year the 13th Young Communist League Congress called on girls to go to the virgin land regions. Now, on the threshold of the new seven-year period, your help is especially needed. Come, dear friends, to do great deeds. . . . The highly fertile virgin lands must give even more bounteously of their riches in the new Seven-year Plan. Girls are working gloriously with the men on the virgin lands. They have taken the initial adversities in their stride; like everyone else, they have slept in tents on which the rain was pouring down and have struggled fearlessly through snowstorms. Together with the men they have made adobe huts, built the first houses, ploughed the virgin land, and sown, raised, and harvested grain. Among those who have been awarded orders and medals are tractor drivers, tractor-drawn implement operators and the glorious brigade house-keepers who also do the combine operators' laundry and prepare borsch for our wonderful fellows. Work on the virgin lands will become the finest school of life. The virgin lands await enthusiastic, patriotic young women!

Extract from Komsolskoya (*the Young Communist League newspaper*).

Two major coalfields are located in this region. The Kuznetsk Coal Basin produces about 16% of all Soviet coal, and about 30% of the total coking coal output. Nearly half of the Kuznetsk coal is used locally, the rest being sent to the Moscow, Urals, Volga and Middle Asian industrial centres. The Karaganda field produces about 5% of the country's coal; most of this is sent to industrial cities in the Urals and Middle Asia. (38) Draw a simple sketch-map to illustrate the facts given above.

Very rich iron ore deposits are located in Northern Kazakhstan. Gornaya Shoria ores supply nearly all the requirements of iron and steel plants in the Kuznetsk Basin. Those of Atasuskiy supply the new steel town of Temir Tau. The Kustanay ore deposits are low grade but very extensive. They supply iron and steel plants in Karaganda and Orsk-Khalilovo (in the South Urals).

WESTERN SIBERIA AND NORTHERN KAZAKHSTAN

ARCTIC CIRCLE

(G) Berezovo

Bitterly Cold Winters
Permanently Frozen Subsoil
WEST
Ob
SIBERIAN
PLAIN

Sverdlovsk
Irtysh
Tobol
Vast Regions of Marsh in Spring & Summer
KUZNETSK BASIN

BASHKIRIA URALS
Magnitogorsk Kustanay (I)
Kurgan
Petropavlovsk
Omsk
TRANS-SIBERIAN RAILWAY
Novosibirsk (DAM)
Kemerovo
Ishim

Orsk
Arid
Steppe
Barnaul
Novokuznetsk (I)

(Ph) Mugodzhar
TURGAY (A)
Temir Tau
Semipalatinsk
Gornaya-Shoriya

LOWLANDS
Karaganda
(R) Leninogorsk

KAZAKH UPLANDS
DAMS

0 100 200 300
km
Dzherkazgan
(M)(C) (I)(C) Atasuskiy

- ● COAL
- (I) IRON ORE
- (A) ALUMINIUM
- (C) COPPER
- (R) PYRITES
- (M) MANGANESE
- (Ph) PHOSPHATE
- (G) NATURAL GAS
- ▨ Main Settled Farm Region

The map on page 231 shows that there are important non-ferrous ore deposits in Western Siberia and Northern Kazakhstan, in addition to large coal- and ironfields. Kazakhstan contains about one-half of the country's reserves of copper, lead and zinc, and large deposits of bauxite (which yields aluminium) are being opened up in the Turgay Lowlands. This mineral wealth forms the basis of metal-working industries in Omsk, Karaganda, Novosibirsk and the Kuznetsk Basin. Other industries in the region are summarised on the left. Possibilities of expanding both the industry and the agriculture of Siberia are studied at the important scientific research centre of Academgorodok, near Novosibirsk.

EASTERN SIBERIA and the FAR EAST cover nearly 8 m. km², i.e. the region is 32 times larger than Great Britain. Yet this part of the Soviet Union contains a mere 11 million inhabitants, compared with 55 million in our own tiny country. Moreover, the map on page 202 shows that most people in the eastern U.S.S.R. live in a narrow, broken belt in the extreme south, most of the north being virtually uninhabited. The climate figures below show why. (*39*) Draw temperature charts on the same axes for Verkhoyansk and Irkutsk and state, for each of these towns, how many months have an average temperature below freezing point. (*40*) State the total mean annual rainfall for each of these towns, and describe its seasonal distribution. In addition to the extreme and semi-arid climate, other hindrances to settlement in the eastern U.S.S.R. are indicated on the map. (*41*) Write them down. The 'pockets' of population shown on the map on page 202 are located mainly in isolated lowland basins and river valleys, where a comparatively mild summer climate and better soils permit cultivation. Grain, vegetables, sugar-beet, various fruits and oriental crops such as soybeans are produced and there is some dairying in the moister districts, e.g. around Vladivostok. The more arid steppe and mountain country is used for sheep grazing, reindeer herding, trapping and fishing. Many farmers find part-time employment in mining, lumbering and fishing enterprises. The region is still only about one-third self-sufficient for food.

		J.	F.	M.	A.	M.	J.	J.	A.	S.	O.	N.	D.
Verkhoyansk	°C	−50	−44	−30	−13	2	12	15	11	2	−14	−37	−47
	Rf. mm	5	3	0	3	5	13	20	23	5	5	5	5
Irkutsk	°C	−21	−18	−10	1	8	14	17	15	8	0	−11	−18
	Rf. mm	15	13	10	15	30	58	73	60	40	18	15	20

The region contains in widely scattered fields (*see map*) about 70% of the coal reserves of the U.S.S.R.; but only the Cherem-khovo, Kansk-Achinsk, Minusinsk and Bureya fields have significant workings, yielding about 12% of the country's total coal output. All this coal is used locally, mostly in chemical works, power stations and as fuel on the Trans-Siberian Railway. Although there are huge deposits of high-grade coking coals in the northern territories, it seems unlikely that they will be utilised in the foreseeable future, on account of their remoteness. The region also has some oil and natural gas (*see map*) and pipelines are being laid to bring in oil from Ufa, on the Volga–Urals oilfields, over 3000 kilometres away. New refineries in Krasonyarsk and Irkutsk, the main industrial cities, will handle this 'imported' petroleum. There are also large h.e.p. stations (and more projected) on the Rivers Yenesei and Angara. It is planned to electrify the entire Trans-Siberian Railway, thus greatly increasing the freight carrying capacity of this 'lifeline' of the Eastern U.S.S.R. Its terminus is Vladivostok, the major Soviet fishing, commercial and naval port

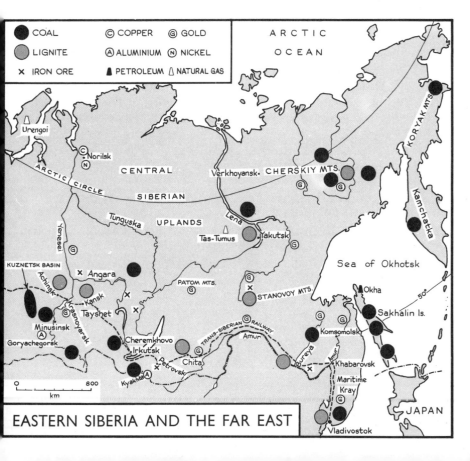

EASTERN SIBERIA AND THE FAR EAST

Iron and Steel and *Engineering.* A small steel plant at Petrovsk (east of L. Baykal) and another at Komsolmsk (in the lower Amur Valley) both use pig-iron brought from the west along the Trans-Siberian Railway. A large new steelworks at Tayshet opened in 1965. This works has been built at a strategic position on the Trans-Siberian Railway, with easy access to several small coal and iron deposits. Initially the Tayshet plant will use coke and coal brought along the Railway from the Kuznetsk Basin. A major steelworks is planned for Krasnoyarsk.

Engineering industries using steel and other metals produced in the region are located in Krasnoyarsk, Irkutsk, Khabarovsk and Komsomolsk.

Non-ferrous mining is important at many scattered locations throughout the Eastern U.S.S.R. The principal minerals obtained are:—*copper* and *nickel*—from Norilsk, in the lower Yenesei Valley; *gold*—dredged from the bed of the middle Yenesei; *aluminium*—refined by h.e.p. at Krasno-yarsk and Irkutsk, from ores obtained at Goryachegorsk and Kyakhta; *diamonds*—dug from mines in many parts of the Central Siberian Uplands (*Map p. 233*).

Lumbering in the region accounts for approximately one-sixth of the U.S.S.R.'s output of sawn timber, and enough pulp and paper to satisfy local needs. All major towns on the Trans-Siberian Railway have saw- and pulp-mills. Other mills are located (*a*) on the Yenesei River, down which wood products are sent in summer for export via the Arctic Ocean and (*b*) on the Pacific seaboard in Maritime Kray and on Sakhalin Island.

Fishing on the Soviet Pacific coast has rapidly expanded in recent years, so that these are now the principal fishing grounds of the U.S.S.R. There are dozens of small fishing ports around the Okhotsk Sea and on the eastern shore of Kamchatka, but Vladivostok is the main base. Salmon and crabs are caught in large numbers and there are whales, walruses and seals in the North Pacific and Arctic waters.

on the Pacific. Vladivostok has engineering, woodworking and fish-processing industries, but its port activities are hampered by ice during much of the winter.

Details of other industrial activities and cities in the region are given in the notes above. (*42*) Summarise all this information (including that about Vladivostok) on a *large*, labelled sketch-map.

Fishing trawlers at their moorings in the great Far Eastern port of Vladivostok. The city is also the Pacific Naval Base of the U.S.S.R. The harbour is kept open by icebreakers all the year round.

Open-cast lignite
excavation near
Cottbus. Lignite is
main power re-
serve of E. Germany.
(See p. 239).

CHAPTER 10

Eastern Europe

(1) EAST GERMANY

The Communist-governed "German Democratic Republic" was established in 1949, with East Berlin as its capital. The boundary between West and East Germany follows no particular physical feature, but in the east the border follows the Oder-Neisse rivers and in the south the crest line of highlands which reach over 1000 m. (*1*) Using an atlas (*a*) name these highlands and (*b*) describe the country's northern boundary

Much of East Germany is a broad undulating lowland, thickly covered in the north and centre by glacial sands and clays: (*2*) where was the source of the ice-sheets which deposited these materials? Winds blowing from these ice-sheets which deposited a thick layer of loess on the southern part of the lowlands (*p. 12*).

The map on page 84 shows that East Germany is essentially the central region of pre-war Germany. Between 1939 and 1949 this region was heavily damaged, first by the ravages of war and then by the dismantling of factories carried out by the Russian occupation forces. Since then, however, redevelopment has taken place under a series of five- and seven-year plans. Industry has been nationalised and concentrates on producing heavy industrial rather than attractive consumer goods. Most of the land is now State-owned and is farmed by collectives. Widespread discontent with these changes added to the large-scale migration of refugees to West Germany referred to on page 106.

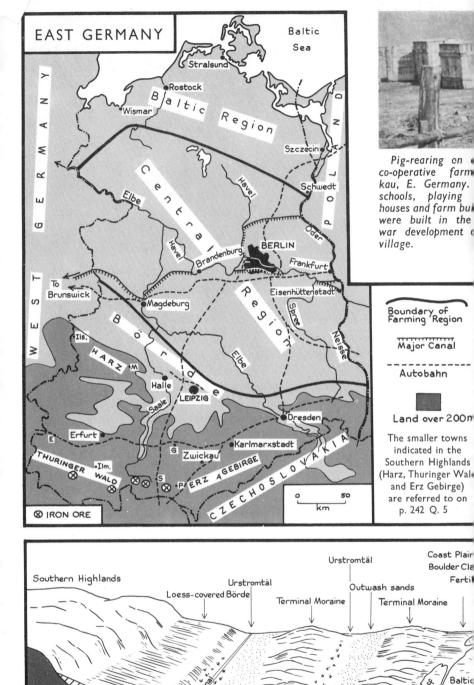

EAST GERMANY

Baltic Sea

Stralsund
Rostock
Wismar

B a l t i c   R e g i o n

Szczecin
Schwedt

C e n t r a l

Elbe
Havel
Havel

BERLIN
Brandenburg
Frankfurt

Oder

P O L A N D

W E S T   G E R M A N Y

To Brunswick

Magdeburg

Eisenhüttenstadt

R e g i o n

B ö r d e

Spree

Neisse

Ils.
HARZ
M

Halle
LEIPZIG

Elbe

Saale

Dresden

E   Erfurt

Karlmarxstadt

THURINGER WALD
Ilm.

G   Zwickau

⊗   ⊗

P   ERZ GEBIRGE

S

C Z E C H O S L O V A K I A

⊗ IRON ORE

0   50
km

Pig-rearing on a
co-operative farm
kau, E. Germany.
schools, playing
houses and farm bu
were built in the
war development o
village.

Boundary of
Farming Region

┬┬┬┬┬┬ Major Canal

- - - - - Autobahn

■ Land over 200m

The smaller towns
indicated in the
Southern Highlands
(Harz, Thuringer Wal
and Erz Gebirge)
are referred to on
p. 242 Q. 5

SIMPLIFIED SECTION ACROSS THE NORTH GERMAN PLAIN

Southern Highlands

Loess-covered Börde

Urstromtäl

Terminal Moraine

Urstromtäl

Outwash sands

Terminal Moraine

Coast Plain
Boulder Cla
Fertil

Canal

OLD HARD ROCKS

Haff

Baltic
Sea

▨ LOESS   ⠿ COARSE GRAVEL & SAND   ▧ FINER SAND   ■ ALLUVIUM

East Germany has a continental climate. In winter the country is swept by bitterly cold winds from Russia; rivers freeze, and snow, though light, lies unthawed for months. Summers are hot and fairly dry, but frequent convectional showers give a summer maximum of rainfall. (*Berlin, Jan. −1°C. (30°F.); July 19°C.; 575 mm mean annual rainfall.*) The Baltic coast plain has a slightly higher rainfall (*why?*) and the Southern Highlands receive up to 1500 mm (*why?*). Except in the Southern Highlands the climate favours arable farming, but it is the *soils* which mainly decide which crops are grown in any particular region. (*3*) Use the notes below to make a labelled sketch-map of farming on the Northern Plain.

FARMING REGIONS OF THE NORTHERN PLAIN

(i) The BALTIC Arable and Dairying Region consists of a sparsely populated undulating lowland. Boulder clays give rise to FAIRLY FERTILE, but HEAVY & POORLY DRAINED LOAM SOILS suitable for arable farming. RYE, POTATOES and SUGAR BEET are widely grown, but FODDER CROPS (hay, roots, green maize, barley, oats and potatoes) predominate. DAIRY CATTLE, BEEF CATTLE and PIGS are reared, the cattle being stall-fed during the long winters. Much BUTTER and CHEESE are produced. PIGFARMS and MARKET GARDENS are very important around the big towns, notably Potsdam, Dresden and Erfurt.

(ii) *The CENTRAL Rye and Potato Region.* Parallel morainic ridges up to 200 m high run east–west across this region. They are separated by broad, shallow valleys (**urstromtäler**) which were cut by melt-waters in glacial times (*p. 11*). The ridges are SANDY and INFERTILE and are mostly covered with PINE FOREST and HEATH. RYE and POTATOES are grown on clayey patches where soils are improved by artificial fertilisers.

The damp alluvial urstromtäler contain many PEAT BOGS, but drained districts support PERMANENT PASTURE for DAIRYING and there is some MARKET GARDENING around the bigger towns.

(iii) *The BORDE Region* contains East Germany's best arable land on FERTILE LOESS SOILS. Over 70% of the land is cultivated. The chief crops are WHEAT, BARLEY and SUGAR-BEET, farmed in rotation with grass and roots. Large numbers of BEEF and DAIRY CATTLE are stall-fed and SHEEP are fed on roots and stubble. There is a big demand for meat and milk in the many nearby industrial towns.

East Germany is not a rich farming country. Even though 47% of the land surface is arable, a mere 7% of the country's soils are really valuable. A further 13% is meadowland or pasture and no less than 27% is forested—mainly the glacial outwash gravels and mountain soils. The remaining 13% includes lakes, built-up areas and barren highlands. (4) Show these land-use percentages in a labelled bar diagram.

Farming in East Germany has been much affected by the drastic changes in landholding since 1945. In that year all estates larger than 100 hectares were seized by the Government, to form a Land Fund totalling 2·2 million hectares. About one-third of this land was then given to farmers whose plots were too small, and the remainder to landless labourers. The mass of new farms soon proved too small to be efficient and so a drive towards collectivization began in 1952. Today more than 90% of crop-land is parcelled into collectives with an average size of about 400 hectares. The Government maintains that each collective has been formed by the voluntary amalgamation of the land owned by individual peasant farmers, but much official persuasion and even coercion has been necessary to get them to band together. Thus the patchwork quilt of individually owned strips and patches has given way to very big fields cultivated in common. In some villages blocks of apartments have been built to house employees from the neighbouring collectives. In spite of such building and a big increase in farm mechanization (the number of farm tractors, for example, jumped from 14 500 to 150 000 between 1950 and 1970) the change in landholding has *not* resulted in any notable increase in crop yields.

" It is evident that East German agriculture has failed to respond either to the territorial reorganization of farms or to the large invest-ment that has been made in it. The reason probably lies in the lack of motivation of the farm-worker; in the flight of the younger and more energetic to the cities or to the West, and the predominantly un-trained and heavily female labour force that is at present available. External factors, notably the weather and the failure of the chemical industries to provide adequate supplies of certain fertilizers, have no doubt contributed greatly to the poor performance of East German agriculture.''*

INDUSTRY

East Germany is short of coal and iron ore, but has vast reserves of lignite and potash. Much coal, iron and steel came formerly from the Ruhr, but since 1945 Poland, Czechoslovakia and Russia have become the chief suppliers. (*For what political reason?*)

* Norman J. G. Pounds, *Eastern Europe*, Longmans, p. 251.

Table A: MAIN INDUSTRIES OF EAST GERMANY	
Main pre-war industries	Communist-planned post-war industries
Chemicals, Textiles, Textile Machinery, Electrical Engineering, Optical and Precision Instruments.	Power Generating and Mining Equipment, Heavy Machine Tools, Structural Steel, Shipbuilding.
Since 1958 the pre-war industries have regained much of their former importance.	

Table B (i) SOME PRODUCT MADE FROM COAL AND LIGNITE

Ammonia (for Nitrogen Fertiliser); Drugs, Plastics, Synthetic Fibres.

(ii) SOME PRODUCTS USIN POTASH IN THEIR MANUFACTURE

Medicines, Soaps, Paints, Explosives, Bleaches, Matche

The type of industry in E. Germany has also changed: Table A shows the increased emphasis on heavy engineering.

(i) *The Halle-Leipzig industrial region* is based on:—

(*a*) *Lignite.* The huge local deposits occur in beds 100–135 metres thick and are worked by highly mechanized open-cast methods. (*See photo on page 235.*) The 260 million tonnes of lignite which are extracted each year represent 36% of the total world production of this mineral. Since it is not economical to transport this poor-quality fuel over long distances, it is mostly consumed on the coalfield. Lignite 'briquettes' are burnt in gigantic power stations built alongside the quarries, 90% of the total East German electrical output being obtained in this manner. Lignite is also used as a raw material in the region's important chemical industries.

(*b*) *Potash and Rock Salt.* The potash beds are worked open-cast and salt is obtained from deep mines in the foothills of the Harz Mountains. Both minerals are used as chemical raw materials: details of the products obtained from lignite and potash are given in Table B, p. 239 (*III: 224*).

Altogether the chemical industries in this region account for 17% of the total manufactures of East Germany and give employment to 10% of the country's working population. Five major chemical works in and around Halle form the nucleus of the

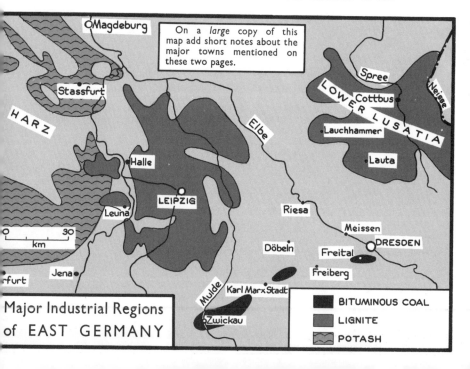

On a *large* copy of this map add short notes about the major towns mentioned on these two pages.

Major Industrial Regions of EAST GERMANY

■ BITUMINOUS COAL
■ LIGNITE
〜 POTASH

industry which is scheduled for further expansion even though it is already the largest in the world. This is because chemicals occupies a vital position in the planned economic future of the country: other industries such as metallurgy and mechanical engineering 'feed' the chemical industry with tools and equipment, while textiles, agriculture and the export trade all depend upon its products. The main towns and industries of this region are:—

Leipzig (596 000), which grew up as a route centre. Its many industries include electrical engineering, machine tools, chemical equipment and textiles. The annual Leipzig Trade Fair commands a world-wide reputation.

Halle (276 000), situated in a lignite and rock salt mining area. Chemicals, machine tools and vehicles are its main products.

Magdeburg (268 000), on a major road and rail crossing point of the Elbe, can be reached by 1000-tonne barges. The river route to the sea is less important than formerly (*Why? See map, p. 90*), but the Elbe links the two largest lignite fields and therefore carries a great volume of traffic within East Germany. The main cargoes are lignite, briquettes, coke, ores, potash, chemicals and iron and steel goods, and Magdeburg is the main focal point of barge movements. This ancient city has therefore become a leading inland port and one of East Germany's main centres of the heavy engineering and chemical industries. (4) Suggest how Magdeburg's position has contributed greatly to its wealth and prosperity as a market centre and as the location of food-processing industries such as sugar-refining. (*Hint: see map, p. 239 and table p. 237.*)

Leuna, the main chemical town, producing polythene, plastics,

The Walter Ulbricht Chemical Works at Bitterfeld

synthetic fibres and petrol. *Jena*, famous for the Carl Zeiss optical works. *Erfurt*, with engineering, footwear and clothing factories.

(ii) *Lower Lusatia* is a rapidly developing mining centre, containing 60% of the country's reserves of lignite. Annual production is now well over 100 m. tonnes (as compared with 60 m. tonnes in 1953). Lusatian lignite is unsuitable for the manufacture of chemicals and is mainly converted into much-needed electricity. Lauchhammer contains the world's first plant for making coke from lignite. Gas from this cokery is piped to the steelworks at Riesa. Local electricity is used to smelt aluminium at Lauta.

(iii) *The Karl Marx Stadt–Zwickau region* possesses the country's only bituminous coalfield. Although this field produces a mere 3 m. tonnes per annum, it supports famous textile and precision light engineering industries. Wool and cotton imported via Hamburg and the Elbe waterway helped Karl Marx Stadt (formerly called Chemnitz) to become the 'Manchester of Germany'. Nowadays the cotton comes from Russia, and home-produced artificial fibres are displacing both wool and cotton. The emigration of skilled labour to West Germany has been felt very keenly in this region of highly technical industries. The main towns and industries of the region are:—

Karl Marx Stadt and *Zwickau*, both producing textiles, textile machinery, machine tools, vehicles and electrical equipment.

Dresden, a beautiful university town and river port controlling the gap through the mountains to the south. Its industries include precision and electrical engineering and technical glassware production. The famous 'Dresden china' is made at nearby *Meissen*.

Freital, the largest coalmining centre; with *Döbeln*, it produces special steels in electrical furnaces.

Freiberg, with metallurgical works once processed locally mined ores, but now relies on imported raw materials.

The Southern Highlands rise gradually from the plains to form a long, level skyline interrupted by occasional rounded summits over 1000 m high. Built of old, hard rocks, they have poor, thin soils and are thickly forested in conifers (*see page 25*). The high plateaus receive over 1500 mm annual rainfall and contain many peat bogs. Throughout the Middle Ages igneous deposits of lead, silver, copper and iron ore were mined, and metallurgical industries grew up using local water power and charcoal. The miners cleared parts of the forests up to about 700 m and grew crops such as rye, oats and potatoes on their small-holdings. This

Mining and Industries of Southern Highlands Towns	
Iron ore	Schleiz, Ilsenburg
Copper	Mansfeld
Uranium	Gera, Aue
Textiles	Plauen
Glass	Ilmenau
Potash	Eisenach
Vehicles	

farming pattern has changed little, but mining has declined. More important are numerous small craft industries which use local raw materials and traditional skills, e.g. wood carving, toys, watches, pulp and paper making, glass and textiles. (5) Make a *large* copy of the *southern* (Highlands) part of the map on page 236 and add the names of the towns given in the list on the left. Draw symbols to show their industries.

Since 1949 considerable changes have occurred in parts of East Germany which hitherto were traditionally agricultural. Chief among the newly-developing industrial centres are:—

Rostock, which has been made into the country's chief commercial and fishing port. Soft muds in the harbour facilitate dock excavation and a new 12 m deep channel has been dredged to enable the port to take large ocean-going vessels. Rostock also has one of Europe's largest and most modern shipyards.

Schwedt has a large new oil refinery fed by pipeline from the Urals. Output should reach 8 m. tons by 1970, enough to satisfy the entire petroleum demand of East Germany.

Eisenhüttenstadt is a new town of 18 000 people, built around a huge iron and steel industry. Coke from Polish Silesia and iron ore from Russia is imported via the Oder waterway. Finished steel is sent to East Berlin via the Oder-Spree waterway. (*Map p. 236.*)

Brandenburg's new steelworks rely heavily on scrap metal. Products include tractors, steel for East Berlin and ships' parts for Rostock

The country's transport system has also been drastically reorganised along Russian lines. 90% of the traffic is now carried by the railways, which are best developed where they lead east to Poland and Russia or north to the Baltic (*see map*). Waterways leading west, i.e. to Hamburg and the Ruhr, have greatly declined in importance. The East German *autobahn* network also carries considerably more traffic than the canals.

The map shows that *East Berlin* (1 074 000) has retained its importance as a route centre by the construction of by-pass canal and rail routes around West Berlin. Factories in the city have been rebuilt since the complete devastation of 1945, so that East Berlin—despite a shabby appearance—is now the largest single industrial centre in East Germany. The city's industries are mainly concerned with electrical and mechanical engineering, but clothing, food processing and other consumer goods products are

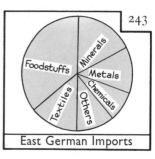

East German Imports

also important. (6) What exactly is meant by the term 'consumer goods'? Why do such industries tend to be concentrated in very large cities?

Despite many geographical disadvantages East Germany now ranks fifth amongst the industrial nations of Europe. In recent years the country's industrial pattern has been varied by the addition of iron and steel production to the traditional highly specialized light industries. This has enabled East Germany to become the largest exporter of machinery within Comecon (p. 31). Most East German trade is conducted with this group of nations. Broadly speaking, imported raw materials are paid for by exports of chemicals and manufactured goods. Despite attempts to improve agriculture, especially livestock production, over one-third of East Germany's imports consist of foodstuffs. (7) Use the figures in the table to construct a diagram showing the country's export pattern.

The drive to expand industry and overseas trade has been hindered by the distance of nearly 400 kilometres between the main Silesian manufacturing cities and the country's seaports. The map shows how this problem has been minimized by the re-orientation of the East German railway network to give rapid access between Silesia and the Baltic coast. East German industry is also hampered by a serious lack of high quality coal—the small bituminous coalfield is both deep and faulted and only one power station, located in East Berlin, uses this type of fuel. Hydroelectric potential is also very limited. Yet another problem—a scarcity of high-grade iron ore—puts a brake on attempts to expand and modernize the new iron and steel industry. On the other hand the country has an abundance of lignite, mineral salts and substantial quantities of non-ferrous metal ores, especially of lead and zinc. Partly to take advantage of these mineral reserves, great emphasis is placed in the current Five Year Plan on expanding the metallurgical industries, especially shipbuilding and heavy machinery.

EAST GERMANY Exports, %			
Machinery	25	Minerals	8
Chemicals	17	Electrical Goods	7
Vehicles	9	Others	34

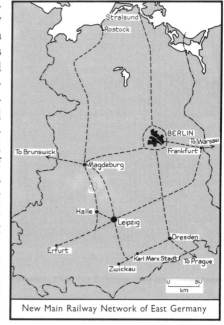

New Main Railway Network of East Germany

(2) POLAND

The word 'Pole' means *plain*, and the Polish people are the plain-dwellers of that part of the North European Plain which lies between East Germany and Russia. In the south the plain is bounded by the Sudeten and Carpathian Mountains, with the Moravian Gate (*see map*) linking the country to Czechoslovakia and the Danube lands. In the east and west this huge lowland lacks well-defined frontiers, and throughout history invading armies have swept across it to conquer and partition the country. Thus between 1795 and 1919 the Polish State disappeared, but it was re-created after the First World War. Further territorial changes were made in 1946 in an attempt to establish more suitable boundaries. In the east 198 000 km² of mainly marsh and forest land, largely populated by 'White Russians', were ceded to Russia, but in the west and north Poland gained 114 000 km² of former German territory. These changes gave the country a substantial coastline with three large ports, much valuable farmland and most of the Upper Silesia coalfield (*see map*).

Communist Poland was established in 1946 when the battered country emerged from the Second World War. In that war Poland had been the scene of the greatest horrors and crimes ever perpetrated against humanity. Six million Poles were killed, many of them in Nazi 'death camps' such as Auschwitz. Warsaw, the capital, lay in ruins and one-third of the country's houses and factories had been demolished. Farmland had been devastated and there was a serious shortage of machines and livestock. In 1945 seven million German civilians were expelled across the new western frontier on the Oder–Neisse. Priority in reconstruction was given to building up heavy industry and reallocating farmland to the peasants. Industry was nationalised, but attempts to 'collectivise' agriculture were abandoned in 1956 after proving very unpopular (*see also p.* 261). Redevelopment is now well advanced, but Poland still lags behind most West European countries.

(8) (*a*) State the mean annual temperature range in each of these towns and explain the differences (see p. 14 if necessary); (*b*) explain why they receive most of their rain in summer (p. 16); (*c*) state what form much of the precipitation would take in winter; (*d*) state Warsaw's total summer rainfall; (*e*) write a short description of Warsaw's climate.

Place	Latitude	Jan. Temp.	July Temp.	Annual Rainfall	Wettest Month	% of Rain in Winter Months	No. of Months with Average Temp. < 0°C.
Berlin	53° N.	−1°C.	19°C.	575 mm	July	43	1
Warsaw	**52° N.**	**−3° C.**	**19°C.**	550 mm	**July**	43	3
Kiev	50° N.	−6°C.	20°C.	525 mm	July	34	4

Warsaw, devastated by bombing in 1940, has been completely rebuilt. Here we see part of Warsaw Polytechnic and the Palace of Science and Culture.

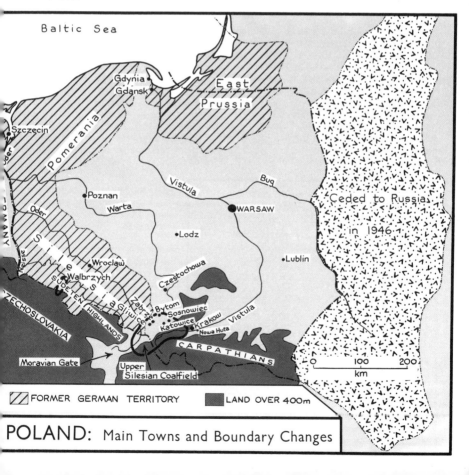

POLAND: Main Towns and Boundary Changes

The physical geography of Poland, and as a result its agriculture, are very similar to those of East Germany. The plains are thickly covered with glacial deposits, including extensive patches of loess in the south. Farms vary in size from 5 to 7 ha, except for a few large State farms located in the former German territories. The main land-use regions of Poland are shown in the map. (9) Label a *large* copy from the following notes:—

1. *The Baltic Lakes Plateau* is bordered in the north by a coast fringed by lagoons, marshes and sand-spits. This coast region contains large areas which are infertile because of light sandy or heavy clay soils, but the rainfall is relatively heavy and there is some dairy farming. Except around the three main ports (*see overleaf*) the only notable signs of habitation are a few small fishing villages and a developing sea-side resort at Kolobrzeg. Inland stretches a region of hummocky boulder clay, terminating about 160 kilometres to the south in morainic ridges which in places are over 300 m high. The moraine country includes thousands of lake-filled hollows alternating with higher patches of heath and forest and during the summer months it becomes a ' Mecca ' for thousands of canoers and campers. Most of the permanent residents live in widely spaced villages located in woodland clearings and are mainly employed in lumber and woodworking industries. The area of farmland has been increased recently by large-scale drainage and soil improvement schemes. Dairying and pig-rearing predominate, and some rye, potatoes and sugar-beet are grown on the lighter, sandier soils.

2. *The Central Plain* consists of flat sheets of boulder clay interspersed with huge patches of wind-blown glacial sand: one notable sand-dune area to the NW of Warsaw covers 3200 km² of the Vistula flood plain. Marram grass and pine trees have been planted to stabilize the sand and prevent it being blown over and damaging other more fertile soils. In the east huge tracts of bog merge into the Pripet Marshes of Russia. Despite a lack of fertile soil and the short growing season, about 70% of the land is cultivated, rye, oats and potatoes being the main crops. Villages in this region are closely packed, each settlement being surrounded by great open fields carved into thousands of elongated strips of land. The villagers own and cultivate various isolated strips, an inefficient pattern of land-holding which has survived since medieval times. Despite the arrival of Communism the amalgamation of strips into farm holdings is progressing very slowly. (*See also p. 261.*)

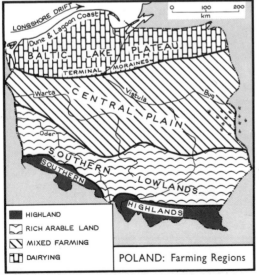

HIGHLAND
RICH ARABLE LAND
MIXED FARMING
DAIRYING

POLAND: Farming Regions

3. *The Southern Lowlands* consist of the plains of the mid-Oder valley and the low plateau country farther east. Much of this lowland is covered with loess and it is the principal farming region of Poland. The loess soils are rich, light, easy to till and give high yields of wheat, barley, sugar-beet and potatoes. Farms tend to be small, about 80% of them covering less than 10 hectares apiece. Poland remains the least collectivized of all East European countries and 85% of the farms are still in private hands. The peasant pro-

prietors remain largely self-sufficient and so few crops are sold on the open market. In places old, hard rocks protrude through the loess, giving rise to patches of forest and moorland.

4. *The Southern Uplands.* Here the rounded summits of the Sudeten Mountains contrast markedly with the loftier peaks of the Carpathians, which reach nearly 3000 m in the High Tatra. The lower slopes of these uplands are heavily forested with beech, spruce and firs, but above the tree-line lie hill pastures and bleak, desolate moorlands. Lumbering and pastoral farming with mountain-valley transhumance are the traditional occupations, but there is a growing tourist industry in the Carpathians. Tourists are especially attracted by the traditional folk-ways which survive in the isolated mountain villages.

Industry. Poland's annual coal output of about 120 m. tonnes has been achieved by intensive working of the Silesian coalfield (*map p. 245*). Since 1946 the Upper Silesian field, which may possess greater reserves than the Ruhr, has been almost entirely within Poland. 90% of the country's coal is produced here from easily mined seams less than 30 m down. Half is exported, mainly to Scandinavia and Eastern Europe. Jurassic iron ore mined near Czestochowa supplies about 12% of the country's needs, but large supplementary supplies are imported from Russia and Sweden. Upper Silesia is the most densely settled and most highly urbanized part of the country, more than 1½ million people being packed into the conurbation shown on the map. This conurbation has evolved since the late 18th century, when the coal-mining and iron smelting industries were encouraged by the then ruling Prussian government. Small villages on or close to the exposed coalfield grew steadily into large industrial cities. More recently these cities have virtually fused together to form a continuous industrial landscape of factories, huge blocks of flats, waste tips and 'flashes' of stagnant water where the ground surface has subsided over coal workings.

" The industrial region . . . consists of about a dozen cities. Around and between these are grouped a number of smaller towns, most of them primarily industrial, some serving as dormitories for the factory workers. In the whole area only two towns, Gliwice and Bytom, are older than the nineteenth century. The rest have mushroomed with the growth of industry: unplanned, ugly, insanitary. Coal workings burrow beneath them, and the spoil heaps from the mines and smelters encroach on them. Wherever there are no buildings, it is usually because the danger of subsidence makes it unsafe to construct them. The industrial plans of Poland make it necessary that these towns should grow; the supply of water and of other services and utilities makes it difficult for them to do so, and the shortage of land makes it almost impossible. The planning of the Upper Silesian industrial region has faced greater problems than that of most other industrial areas in Europe, and it was obliged to resort to more drastic means. The urbanized area must expand its industries, many of which are tied to particular sites, but more and more of the workers are obliged to live outside the area. Satellite or dormitory towns are being built each made up of huge apartment blocks for industrial workers who travel daily by train, bus and bicycle to the factories and mines within the urban-industrial complex. Nowy Tychy, with a population of over 50 000, is the largest of such proletarian towns. The water needs of this region are immense and are growing daily, with the expansion of branches of industries, such as steel-making, which are heavy consumers of water. Local sources have long since been exhausted; the Vistula itself, 40 kilometres from the industrial region has been dammed at Goczalkowice to supply it; the Beskidy have been tapped, and now attention is turning to more distant sources in the Tatry.''*

The Lower Silesian coalfield is centred on Walbrzych in the Sudeten foothills. Annual production is only 3·2 m. tonnes. In the rich farmland to the north is the river port of Wroclaw, Silesia's chief town. Its industries include flour-milling, sugar-refining and the manufacture of chemicals, locomotives and woollens.

Poland's other major industrial centres badly need developing. Roads are often poorly surfaced and rivers, although greatly used, need much attention to improve their navigability. The railways, however, are being modernised; the Warsaw–Czestochowa line, for example, is now electrified.

In many respects Poland retains its rural and peasant traditions, but recently some very big strides have been made in industrial development. As a result of a series of Five Year Plans about 70% of the country's manufactures now come from plants and factories built since 1954, and industry's contribution to the national economy has risen from 53% to 82%. Expansion is favoured by the existence of a young, vigorous working population—one-third of all Poles are under sixteen—but there are also hindrances due to

(i) an almost complete reliance on imported Russian petroleum —although there have been recent discoveries of natural gas in the Carpathian foothills in S.E. Poland; and

(ii) the crowding and lack of space for industrial expansion in the main industrial region of Upper Silesia (*see also p. 247*). The principal growth industries being fostered in the current Five Year Plan are chemicals, especially petro-chemicals at Plock ((*10*) *Why there? see map, p. 28*); metallurgy, e.g. lead and zinc smelting

* Norman J. G. Pounds, op. cit., p. 356.

Warsaw (1 300 000) grew up at the main crossing point of the Vistula, to become Poland's foremost road and rail focus. Spectacular growth in the late C19th followed its deliberate choice as a textile manufacturing centre, mainly using imported cotton. 80% destroyed during the 1939–1945 War, the Old City was rebuilt in its former style as a symbol of Polish nationalism. Its new buildings include Government offices, banks, business houses and numerous factories. Textiles, processed foods, books, electrical equipment, iron and steel goods and vehicles are the city's main products. Further expansion is hampered by a lack of local raw materials and by woefully inadequate water supplies.

Lodz (746 000) is Poland's second largest town and the country's 'textile capital'.* It produces cottons, woollens and clothing as well as electrical goods and chemicals.

Krakow (526 000), once the capital of Poland, is a rail centre on the route through the 'Moravian Gate'. Its industries include metallurgy, engineering, paper-making, flour-milling and chemical manufacture, using local salt deposits. Six miles from Krakow is the new steel town of *Nowa Huta*, where iron ore from the Ukraine and coke from Silesia and Moravia are used to produce about 1·5 million tonnes of steel annually. Nowa Huta was built to absorb workers from the overpopulated rural region of S.E. Poland. Similar deliberate expansion of industry to relieve rural unemployment is taking place in *Tarnów* (chemicals and engineering) *Rzeszów* (engineering and building materials), and *Torun* (textiles).

Poznan (441 000) has industries closely connected with the town's position in the centre of rich farmlands. Agricultural machinery, fertilizers, flour and beer are produced, and there is an annual International Trade Fair.

The MAIN PORTS of Poland, in order of importance, are:—

1. *Szczecin* (315 000). Products: Iron and steel, ships, paper, synthetic fibres, precision instruments.

2. *Gdansk* (324 000). Products: Ships (see photo), engineering and electrical goods, chemicals, paper, cellulose, woollens, dairy produce.

3. *Gdynia* (168 000). Products: Ships, engineering goods. The port, which is entirely artificial, was constructed in 1925. It is specially equipped to handle bulk cargo carriers, but has suffered recently by a catastrophic drop in coal exports. The main cargoes handled are iron ore, coal, grain and lumber. (9a) What advantage has Szczecin over Gdansk in trading with Silesia, as regards transport?

---

* New textile mills have recently been established in the largely rural region of western Poland. The main new centres are at Zagan and Torun, the latter town now containing Poland's largest textile factory, with over 6000 employees. Wool is imported via Gdynia and there is an increasing use of man-made fibres.

(See also p. 245)

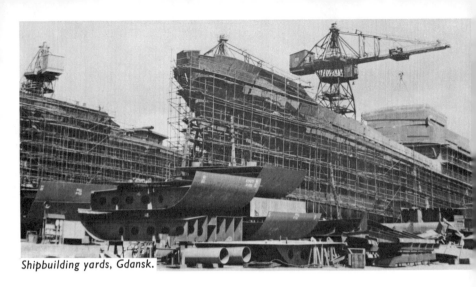

Shipbuilding yards, Gdansk.

from local ores near Bytom, and steel-making in Upper Silesia; and metal fabricating, mostly in the engineering works of Upper Silesia. In addition new steel-making centres are being expanded near the Jurassic iron ore at Czectochowa, at Nowa Huta (*see p. 245*) and in Warsaw.

The immense reserves of low-cost Silesian coal remain the key factor in Poland's industrial development, for 90% of the country's electrical power comes from coal-fired power stations. It is planned to expand production from the present 130 million tonnes to 160 million tonnes per annum by 1975. The collieries are being automated to improve productivity and take advantage of coalseams which in places are 10 m thick; operations such as cutting, loading and even roof supporting are effected by remote control and closed circuit television. The completely automated Jan colliery at Katowice is one of the industrial show-pieces of Eastern Europe. In addition to hard coal, Poland possesses enormous reserves of lignite. These are as yet little exploited, but opencast workings have begun close to the East German boundary and at Konin. Most of the present output of 25 million tonnes p.a. is used to generate electricity in power stations adjacent to the workings, as for example at Turoszów.

FOREIGN TRADE OF POLAND	
Main Exports	Main Imports
Machine tools	Petroleum
Textile machinery	Machinery
Rolling stock	Iron ore
Coal	Chemicals
Foodstuffs, e.g.	Grain
bacon, butter,	Fodder
canned meat	Cotton

Poland has become one of the world's leading shipbuilders and a substantial exporter of ships and engineering products. (*See table*). Trade is mostly conducted within Comecon, but commercial links with Western European countries are increasing and Polish ham, bacon and eggs are well known in Britain.

(3) CZECHOSLOVAKIA

The former Austro-Hungarian Empire collapsed in 1919 and two of the newly-freed national groups, the Czechs and Slovaks, united to form the state of Czechoslovakia. This state stretches for nearly 800 km along the mountainous heart of Central Europe. It includes the very important lowlands of the 'Moravian Gate', linking the North European Plain with the Danube Basin lowlands (*see map*). However, Czechoslovakia has no direct access to the sea. (*11a*) Name the countries (labelled 1–6 on the map) which enclose it. (*11b*) Which two rivers give access for overseas trade via (*a*) the North Sea and (*b*) the Black Sea?

Before 1939 the printing on Czechoslovakian banknotes was in six languages, an indication of the number of national groups within the country. Skilled German workers, Magyar farmers and Jewish merchants all formed important sections of the population. During the Second World War, however, the Jews were killed or deported and most Germans and Magyars were expelled in 1945, after Russian armies had liberated the country from German occupation. In 1948 Czechoslovakia became Communist and the country is now a member of Comecon.

	Jan. Temp.	July Temp.	Annual Rainfall	Midsummer (J., Jy., A.) Rainfall	Winter Snow
Prague (221 m)	−1°C.	20°C.	483 mm	40% of total	Light

These figures reflect the Central European climate of the lowland farming country; but in such a mountainous land altitude and aspect have marked effects. High, exposed westerly slopes receive a precipitation of up to 2000 mm p.a., including heavy winter snowfall. Snow lies all the year on the highest ground, but not on south-facing slopes. (*12*) How much rain do the lowlands receive in mid-summer? (*13*) How is this rain likely to be formed? (*14*) Why is it so cold in winter?

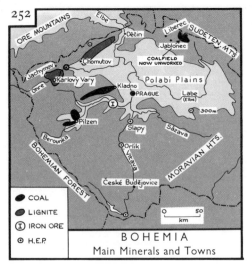

BOHEMIA
Main Minerals and Towns

COAL
LIGNITE
IRON ORE
H.E.P.

75% of the Czech population live in the historic province of Bohemia. The centre of this saucer-shaped region is drained by the Upper Elbe (Labe) and its tributaries. In the Sudetens and Bohemian Forest the surrounding mountainous rim reaches an average height of 700 m–900 m, but in the north, between the Sudetens and the Ore Mountains, it is sharply broken by the Elbe Gorge with its spectacular sandstone cliffs. Throughout the Bohemian highlands the skyline is broken by rounded granite domes, and the mountain slopes are trenched by hundreds of steep-sided valleys.

Of Czecho-Slovakia's three provinces, Bohemia, Moravia and Slovakia, the first is by far the most important.

BOHEMIA. The Mountain Rim is densely forested with conifers and there are many traditional timber-using industries, e.g. making furniture, clocks, toys and musical instruments. These goods were formerly hand-made by craftsmen in their own homes, but now they are mostly produced in small factories using h.e.p. Many sheltered valleys in the forested region have long been cleared of trees and now contain small collective farms and mining villages. The farmers grow rye and potatoes, and graze cattle on meadowland and upland pastures. Hay is the chief fodder in the long snowbound winters.

Mining has been carried on here for many centuries, particularly in the Ore Mountains, where in the Middle Ages German immigrants exploited deposits of silver, lead, zinc and copper. Some of these ancient mines are still productive. There are also local industries based on workings of kaolin, pitchblende, sand and iron ore. (*15*) Which of these minerals provide raw material for the following manufactures:—*porcelain* at Karlovy; *glass* at Jablonec; *uranium* (for export to the U.S.S.R.) at Jachymov?

Liberec is an important textile centre and Karlovy Vary, with its mineral springs, is a famous spa-town.

Central Bohemia. (i) *Farming.* The densely populated Polabi Plains (*see map*) are largely covered with loess and alluvium. Together with the valleys of the Vltava, Berounka and Ohre they form the province's best farmlands. Until 1948 there was still much peasant cultivation on medieval open-fields, but now collective farming is almost universal. Wheat, barley, sugar-beet and

Modern grain silos on collective farm at Zábědov, Bohemia.

potatoes occupy 75% of the land, but flax and hops are also grown on the better soils. To the south the land is higher and cattle rearing is important on permanent pastures.

Central Bohemia. (ii) *Industry.* Central Bohemia is the country's chief industrial region. Local coal and lignite are used in generating electricity. Together with small deposits of iron ore, they form the basis of important iron and steel, engineering and chemical industries. Supplementary supplies of coal from Poland and of iron ore from the U.S.S.R. are imported. Two large h.e.p. plants have recently been completed on the Vltava south of Prague, at Orlik and Slapy.

TOWNS AND INDUSTRIES OF BOHEMIA

Prague (1 030 000) the capital of Czechoslovakia, grew up at a fording (and later bridging) point on the Vltava. It is at the limit of navigation for 500-tonne barges and is an important road and rail centre (qu. (23), p. 257). Manufactures include steel, heavy and light engineering, processed foods, paper, clothing and leather goods.

Pilzen (143 000) is Bohemia's second major industrial town, with a huge nationalised iron and steel plant and important engineering works. The local coalfield produces 3 m. tonnes p.a. but, like that at Kladno, is nearing exhaustion. Local barley and hops supply Pilzen's world-famous breweries. Food processing is important.

Kladno (50 000) has a small coalfield (c. 2 m. tonnes p.a.) supplying fuel for the thermal power stations in Prague. The town also has a light engineering industry.

The *Karlovy Vary* and *Chomutov-Decin* lignite fields produce about 50 m. tonnes p.a. They support a dense population employed in heavy engineering and chemical manufacturing, and in thermal power stations burning lignite.

Textiles, paper, matches and leather goods are manufactured in many Bohemian towns, as are alcohol and starch (from locally-grown potatoes and maize) and sugar.

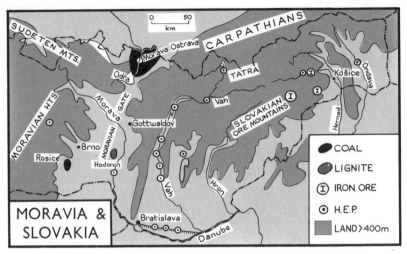

MORAVIA & SLOVAKIA

MORAVIA and *SLOVAKIA*. *Moravia* occupies the central lowland 'corridor' which rises gently to the Carpathian foothills and Moravian Heights at its eastern and western edges respectively. Its open rolling plains are widest in the south, where they reach the Danube, and narrow considerably in the north to form the 'Moravian Gate' between the Sudeten Mountains and the Carpathians. ((*17*) *Which plains does this 'gateway' connect?*)

The central lowlands form valuable farmlands, especially where large expanses of loess occur. Collective farms occupy over 70% of the area, of which 80% in all is arable. Wheat, barley, sugar-beet, potatoes, hops and tobacco are grown, with maize in the south (*18*) *why there?*). Cattle and sheep are reared in the foothills, and in the Carpathians fruit and vines are important on south-facing slopes.

This pattern of farming is repeated throughout the Danube lowlands of Southern Slovakia, where the river is still in places liable to flood despite the construction of drainage channels and embankments. The main flood danger is in spring. ((*19*) *Why?*)

Most of *Slovakia* forms a striking contrast to the efficient farming of the lowlands. The thinly-populated Carpathian Mountains dominate the landscape, and the life of the people is simple and even primitive. In the north the old hard rocks of the High Tatra reach over 2900 m and contain many glaciated pyramidical peaks and arêtes above the tree-line. Elsewhere the mountains are lower, and in the south the Danube lowlands are easily reached *via* many well-defined north–south valleys such as the Vah.

Only rye and potatoes can be grown above 700 m, so farming is mainly confined to these valleys, where barley, sugar-beet and

Boot and shoe factory, Gottwaldov.

potatoes are grown. Sheep and cattle are grazed on the poor pastures of the mountain slopes, but there is little transhumance. Slovakia is the country's chief sheep-rearing region.

Commercial dairy farming is practised in both Moravia and Slovakia, but with a difference. (20) Suggest what this difference is, after deciding which province is the better placed to supply the main cities (*a*) with liquid (fresh) milk or (*b*) with butter and cheese.

Equally contrasting is the industrial development of the two provinces. In northern Moravia the country's largest coalfield produces 18 m. tonnes p.a. and supports several large towns, including Morava Ostrava. Farther south are Brno, the largest town of either province, and Gottwaldov, the 'Northampton' of Czechoslovakia. In Slovakia only Bratislava is of similar importance, as most of the population live in small towns and villages in the valleys. Here the main industries make use of local timber and produce lumber, pulp, paper, cellulose and matches. Textile manufacturing is also important. In recent years the output of h.e.p. has rapidly increased, chiefly on the Vah, which already has seven completed power stations transmitting electricity to industrial towns farther west.

MAIN TOWNS OF MORAVIA AND SLOVAKIA

Morava Ostrava (265 000) is situated on a coalfield and is important for iron- and steel-making. Chemicals are also produced, using by-products from the numerous coke ovens.

Brno (335 000) has a small coalfield and 'imports' electricity from Bohemia. It is noted chiefly for its heavy engineering products and textiles.

Gottwaldov is a modern industrial town specialising in leather goods and plastics. Its Baťa shoe factories are reputed to be the largest in the world.

Bratislava (280 000) is a river port on the Danube and Slovakia's largest town. Clothes, leather goods and explosives are manufactured here.

Above: *Technicians of the Bohemian Metals Research Institute, testing new aluminium alloys for strain under very high temperatures. The aluminium is for use in the Slovakian nuclear power station in Jaselske Bohunice.*
Below: *New chemical plant under construction in Bratislava.*

Communications and Trade. The waterways of Czechoslovakia are handicapped by winter freezing, spring flooding and
summer low water, and only the Labe (Elbe) and Vltava are of
any importance. Railways far outstrip waterways as carriers of
freight. The main goods carried are coal, wheat and sugar-beet.
Since 1947 road and rail routes have been improved, especially
where they are linked to the U.S.S.R. ((*21*) *Why?*) There are
plans for a waterway from the Danube port of Bratislava to
Szczecin at the mouth of the Oder.

(*22*) Make a full-page outline of the map on page 241. Shade
the highlands and mark the main rivers. Mark and name Prague,
Pilzen, Cheb, Brno, Morava Ostrava, Bratislava, Košice, Ceske
Budejovice, Vienna, Dresden, Budapest. Using an atlas, draw
the following railways:—

(a) Prague to Cheb	(b) Bratislava to Brno
,, ,, Dresden	,, ,, Morava Ostrava
,, ,, Ceske Budejovice	,, ,, Vienna
,, ,, Brno	,, ,, Budapest
,, ,, Pilzen	,, ,, Košice

(*23*) What can you say about the position, in relation to communications, of (*a*) Prague and (*b*) Bratislava?

(*24*) How does relief affect the railway network of the country?

(*25*) Why does the Vienna–Bratislava–Budapest railway line
not follow the Danube?

Czechoslovakia has developed a fairly well-balanced economy
and is now the most prosperous state in Eastern Europe. Its
population is about 14 millions. The introduction of collective
farming has brought much more machinery to the farms and has
released labour for work in industry. Only about 25% of the
working population is now in agriculture. In manufacturing,
where the emphasis is on heavy industry, the chief problem is a
shortage of fuel and power. Substantial increases have been
made, however, in the output of coal and lignite. H.e.p. is also
being extensively developed, particularly in Slovakia where the
Government is trying to industrialise this hitherto rural region.

Czechoslovakia is slowly increasing its trade with Western
Europe, but 70% is still carried on within Comecon.

Main imports	Main exports
Wheat, vegetable oils, cotton, wool, hides, iron ore, oil, machinery, chemicals.	Sugar, timber, cotton and woollen fabrics, iron and steel goods, boots and shoes.

(*26*) How far do these lists reflect the pattern of farming and manufacturing described
in pages 251–257?

As most people in South-east Europe are peasant farmers, they have been chiefly affected by the changes in agriculture—changes symbolised by these photographs taken in the rolling fields of North-eastern Romania. The peasants on the left, dressed in long homespun tunics are harvesting wheat with sickles. On the right we see workers on a nearby State farm: they are hoeing a bumper crop of maize with some of the 100 000 tractors in which Romanian co-operatives have invested

SOUTH-EAST EUROPE was for long one of the most under-developed and primitive parts of the Continent. For centuries the region was afflicted by warfare and political upheaval, so that its long-suffering peoples—the great majority of them peasant farmers—had little chance to improve their living standards. We noted on page 29 that the Industrial Revolution, which in the 18th and 19th centuries did so much to transform such countries as Great Britain and Germany, made little impact in South-east Europe. Now, under Communist control, great efforts are at last being made to improve the economy by (i) developing industries, especially iron and steel, engineering and food-processing; (ii) increasing power output; (iii) making farming more mechanised and scientific and larger in scale.

The changes in farming are mainly in the use of co-operative and "collective" methods of organization.

"The collective farms were started after World War II. In many regions the land had belonged to large landowners and was at first distributed among the peasants. Then, in order to improve the backward, medieval standard of agriculture and train people, new ways of farming economically had to be found, and collectivisation of the land was introduced by the Rumanian government, which provided in particular, financial aid for mechanised equipment. By now co-operative work and group activity has become a customary form of life for the peasants, who are beginning to find leisure time for sport and education. Frequently we found cultural centres connected with the cooperatives where lectures and dances, exhibitions and meetings are

258

organised. An interesting development is the handicraft cooperatives. In some regions traditional crafts like carving, embroidery or pottery are still practised in the peasants' homes and the products sold through cooperatives. In other areas, larger centres have been created, with the preservation of the old crafts in mind as much as organised production and sales.

"... At Chizatau we found ourselves at the headquarters of a collective farm, a white-washed building neatly labelled in black lettering. Through the open door we went into a hall. There were notice boards on the walls: production graphs and statistics, the members' work records, announcements of coming events—a cinema, a book exhibition, the folk-singers' rehearsal times. A young man came forward to ask what we wanted. He turned out to be the secretary. ... 'We grow a lot of tobacco here and some maize' he said. 'The cooperative has not been going long, people were slow to join. In the beginning only the smaller peasants put their land together; since they could get government loans to buy machinery, they soon did well. Then the taxes got heavier for the private owners, and all the bureaucracy and paper-work involved in buying and selling got more complicated, so in the end they joined the cooperative too. There are over 600 families in Chizatau, and all of them are now members of this collective farm.' He went on to explain that the working hours of each member and the quality of his work are recorded, and payment is made according to what he has done. In the autumn, after the harvest, the profits are shared out, partly in produce, and partly in money realised from the sale of tobacco."*

In the 1950s collective farms of this type were established in all the Communist countries of S.E. Europe, but they have not all been successful. Rural peasants are well known for their conservative outlook. Opposition to new systems of land-ownership has led to their being 'soft-pedalled'—in Yugoslavia, for example, the State owns one-third of the land but it is still mostly farmed by individual peasant families. In time, however, as the older generation retire, the new co-operative methods may become more widely accepted.

* Alberto Tessore, 'Cooperatives in Rumania', *Geographical Magazine*.

The Danube Basin contains the biggest areas of farmland in South-east Europe. These include the Hungarian Plain, other large portions of the Danube flood-plain and the valleys of various tributaries. The rest of South-east Europe consists largely of sparsely-peopled mountains and plateaus, little of which is suited to farming.

(27) From the map overleaf name (a) the main tributaries of the Danube, and the countries through which they flow; (b) the surrounding highlands; (c) the famous "gates" by which the Danube reaches its lower plains course.*

Other important farming areas, outside the Danube Basin, are also noted on the map overleaf. *Bulgaria*, with more than 40% of its land surface under cultivation, is unquestionably the best-endowed farming country in South-east Europe.

Yugoslavia, except in the Sava-Danube plains, consists mainly of mountainous terrain. The dominant rock is an exceptionally pure white limestone, and there are large tracts of barren *karst*. Much of the higher ground is wild, remote country with bare rock pavements cleft by precipitous gorges. The burning drought of the Mediterranean summer adds to the natural aridity of the permeable limestone, making the Dinaric Alps the poorest farming land in the Balkans.

Despite a considerable recent increase in the growth of towns most people in Yugoslavia still live in scattered, often isolated farms, villages and agrarian settlements (*compare Southern Italy,*

* This gorge (*photo p. 13*) is now the site of Europe's biggest h.e.p. scheme outside the U.S.S.R. A great dam near Orşova holds back a lake 130 km long, and lock gates enable sea-going craft of up to 5000 tonnes to reach up-stream as far as Belgrade. Power output from the dam is greater than that from Aswan on the Nile, and the water is used for irrigation in Wallachia and northern Bulgaria.

A Typical Polje in the Yugoslavian Karst

BARREN MOUNTAINS

ROUGH PASTURE

LIMESTONE SCARP

SPRING

SCRUB

VILLAGE

RIVER

CULTIVATED POLJE

BARE ROCK PAVEMENT

The surface of the *karst* is pitted by innumerable solution hollows (*doliny*), the largest of which (*polje*) are hundreds of metres deep and several kilometres across. The floors of the *polje* retain a thin covering of fertile clay, and spring water seeps into the basins from the adjoining uplands as shown in the diagram. Thus the *polje* provide fertile ' oases ' within which farming villages are clustered. Irrigated crops of vegetables, wheat, maize and tobacco are grown, and cattle are raised on fodder grasses. There are many pigs, and scrawny goats and sheep roam the scanty upland pastures.

page 168), and about 57% are engaged in semi-subsistence agriculture. Many farm holdings are extremely small, and there has been little collectivisation.

Albania is also mountainous and has little good farmland. Until recently its crop yields were the lowest in Europe. In an attempt to improve output, most farms have now been collectivised and the total area under cultivation has been greatly extended. Maize is the main crop but (as elsewhere in South-east Europe) the traditional emphasis on cereals is being deliberately altered in favour of industrial crops like cotton, sugar beet, sunflowers and tobacco. The area under irrigation has expanded more than fivefold during the past thirty years, but increases in food output are matched by rapid population growth; hence living standards remain low and there is still much distress in rural districts.

There is no doubt that agriculture in South-east Europe has advanced greatly during the past 25 years. Improvements include the amalgamation of thousands of extremely small peasant holdings into larger, more effectively-worked farms; mechanisation, especially in the substitution of tractors for oxen and horses; drainage schemes; irrigation, notably of high-value crops such as vegetables and cotton; the setting-up of government agencies to speed the delivery of crops from farms to markets, and to encourage wider use of fertilisers and up-to-date farming methods. Except in Poland and Yugoslavia, these advances have gone hand in hand with collectivisation (*see below*) and even in these two countries collectivisation is the government's declared long-term goal.

The Extent of the Collectivisation of Farmland in Eastern Europe

% of total farmland	
East Germany	89·1
Poland	14·0
Czechoslovakia	c. 90·0
Hungary	c. 97·0
Romania	95·0
Yugoslavia	13·0
Bulgaria	c. 95·0
Albania	c. 93·0

"Collectivisation has everywhere brought changes in the physical landscape. Except in Poland and Yugoslavia, the patchwork pattern of small and irregular strips, separated by unseen boundaries, known only to the peasants who cultivate them, has disappeared, and with it much of the variety and colour from the landscape. As a general rule, fields are very large, for combines and large tractors cannot operate in confined spaces. Now, instead of the multi-coloured plots of land, one sees gangs of brightly clothed women working their way across vast, uniform fields.

Collectivisation has also necessitated the erection of new farm buildings. The clustered stables, cow-sheds, barns and equipment parks, generally built of masonry, roofed with tiles and surrounded by paddocks and feed-lots for the animals, make an impressive sight, and in many areas are the dominant feature of the landscape."*

* Norman J. G. Pounds, op. cit., p. 149

THE WESTERN PLAINS OF TRANSYLVANIA

produce much wheat and maize from fertile loess and allu
soils. As elsewhere in Romania, farmers are pressed by gove
ment policy to grow increasing quantities of industrial crops (
sunflowers, *sugar beet*, *hemp*). There is intensive market garden
near towns.

THE HUNGARIAN PLAIN

owing to its continental climate and light rainfall, was
once covered in steppe grassland or *puszta* (*pp. 19 and 23.*)
Much of this is now ploughed
up and most livestock are stall-
fed. Pastoral farming is important today only on poorer
soils (e.g. the sandy grasslands
of Kecskemét and Hortobágy),
damper riverside meadows
and in the Little Alfold. Immense numbers of swill-fed
PIGS are reared near Budapest.
Elsewhere crops predominate,
wheat mainly to the east of the
Tisza, rye mainly on sandy
areas of the Great Alfold and
oats (for horse fodder) in the
Drava Valley. There has been
a large recent increase in
MAIZE, which now covers
25% of all Hungarian cropland,
and is specially important in
the Great Alfold. Irrigated
rice is grown in the valleys of
the Tisza and Körös. Fruit is
important on sandy soils between the Tisza and Danube.
The main fodder crops are
clover and lucerne.

NORTHERN YUGOSLAVIA

Yugoslavia's best farmlands lie in the
lowlands of the Danube and its tributaries. Broad, well-drained river terraces covered with clay, alluvium and
loess are intensively cultivated for wheat,
maize, tobacco, hemp and sugar beet.

Water shortage is a problem on higher ground, for
rainfall is light and the loess soils are highly permeable.
By contrast the lower flood-plains are very swampy
and every spring brings the threat of disastrous floods.
Large drainage schemes have now reclaimed much of
this land, and pioneer settlers have begun to farm it.
There are many collective farms (unusual in Yugoslavia)
on land seized after 1945 from German-speaking and other
'enemies of the People'.

The most productive district is that north of Belgrade (the
Bačka-Banat). It produces 85% of Yugoslavia's SUGAR BEET,
82% of its VEGETABLE OILS, 74% of its VEGETABLE
FIBRES and 50% of its WHEAT and MAIZE.

THE TUNDZHA BASIN

has substantial areas of irrigated I

THE KAZANLAK BASIN

has extensive rose gardens.
petals are used in making the fan
Attar of Roses perfume.

THE TRANSYLVANIAN BASIN

a region of rolling hills and of broad shallow valleys in which
posits of clay, marl, alluvium, loess and volcanic ash yield deep,
rtile soils. The altitude (*300–600 m*) gives cooler and moister
mmers than in other lowland regions of Romania. Hardy crops
e barley, oats and potatoes are grown as well as wheat and maize.

MOLDAVIA

is marshy in the south but
elsewhere very dry. The
whole region is poor and backward, a legacy of its turbulent
past as a natural "corridor"
lying between the Carpathians
and the Black Sea. Cereals,
notably maize, are grown, but
yields are low. Some excellent wines are produced.

WALLACHIA

is Romania's largest and economically most important lowland region. From the foothills of the Transylvanian Alps
the land slopes gently down
from 300 m to a mere 20 m in
the Danube valley. The fertile
loess and alluvium is intensively
cropped with MAIZE, SUN-
FLOWERS and SUGAR BEET;
but along the Danube a wide
belt of marshy land is suitable
only for winter grazing.

Eastern Wallachia has only
about 500 mm mean rainfall,
and irrigation is essential,
especially where the soil is
sandy. Irrigated VEGETABLES
are a speciality near Bucharest.

THE SOFIA BASIN

is a former lake bed, with
alluvial soils on which cereals
and market garden produce
are grown.

THE DANUBE DELTA

is very marshy and has traditionally provided winter
grazing for cattle brought
long distances from the Carpathians. Now the local
fisheries are more important,
but sedges are harvested for
cellulose manufacture.

THE MARITZA VALLEY

is a broad, level, fertile alluvial plain forming the
geographical and economic heart of Bulgaria.
Farming is intensive, with increasing emphasis on
market gardening and on high-value specialist
crops (e.g. tobacco, sunflowers (for vegetable oil),
cotton and sugar beet).

INDUSTRY IN SOUTH-EAST EUROPE, like farming, has expanded greatly since 1945. This has been due in part to the desire of the Balkan peoples to shake off their feudal and largely rural past, and in part to Soviet persuasion, for the U.S.S.R. gains added security from a strong, industrialised "barrier" between itself and the non-Communist "West". The U.S.S.R. also obtains increasing supplies of manufactured goods from its East European satellites.

Romania has the best-developed industries in South-east Europe, chiefly because it also has Europe's largest oilfield (outside the U.S.S.R.) The field at Ploesti has been worked since 1856; but, as the diagram suggests, large-scale exploitation came only with the various Five Year Plans that began in 1948.

The large and expanding petro-chemicals industry is based on six modern, fully-automated refineries. The oil industry receives 20% of all Romanian capital investment, and imports of crude petroleum from the U.S.S.R. and the Middle East are already much greater than home production. Refined petroleum and petro-chemicals are exported, mainly to Comecon countries.

By its demand for such products as drilling gear, storage tanks and pipelines the oil industry has been a great stimulus to engineering. Romania is the second biggest exporter (after the U.S.A.) of refinery and drilling equipment.

Other industries are noted overleaf. The present policy is to increase greatly the output of light engineering and consumer goods, and links with the West are being strengthened, e.g. the French firm of Renault operates a car factory in Arad, using Russian-made gear boxes. Tourism is being strongly encouraged, and two-thirds of Romania's foreign tourists come from non-Communist countries. *Hungary's* drive towards industrialisation is illustrated in the diagram. Though considerable progress has been made, the country is poor in basic minerals and in power resources—hence Hungary's inability to develop its bauxite deposits to the full (*see overleaf*).

Hungary has a long-established interest in the motor industry, for it was a Hungarian, Donat Banki, who invented the carburettor in 1893, and the world's first tricar and two-cylinder front wheel drive car were produced there in 1900–1902. While the emphasis

ROMANIA: Output of Petroleum and Natural Gas

PETROLEUM (million tonnes)	NATURAL GAS (million cubic metres)
1948 — 4	1000
1969 — 13	17000

refinery at Teleajen, near Oloesti, Romania

is still on engineering there is a growing petro-chemicals industry centred on Budapest and Szolnok and based mainly on imported Russian oil. Petroleum supplies 45% of Hungary's total energy requirements.

Bulgaria has acquired practically all its industries since 1947. Though by comparison with, say, Poland these are still puny, the industrial revolution has already transformed Bulgaria's economy. Factory output now exceeds farm products in value, and with a steady migration from the countryside the urban population has doubled to 40% of the total.

This has been made possible by a large increase in foreign trade, especially with the U.S.S.R. and other Comecon countries. Tourism provides a growing link with the West; the widely-advertised Black Sea resorts include a brand new holiday town at Zlatnipjasăci—"Golden Sands".

Yugoslavia remains one of the least industrialised states of Europe, with only 18% of the working population engaged in manufacturing. A principal reason for this has been the grave lack of good coking coal. Rich deposits of metal ores (*see overleaf*) are now being exploited, and industry is expanding.

1950	HUNGARY	1970
50%	Proportion of Population in Industry	70%

ROMANIA

Petroleum. Four main oilfields, recently augmented by wid
separated new finds. Refinery consumption now greatly exc
home production

Petro-chemicals. Rapidly increasing output of (e.g.) synthetic rub
and textile fibres, detergents, fertilizers and pesticides.

Steel. Mainly from Galati (*see below*) or imported from USSR.

HUNGARY

Bauxite in the Bakony-Vértes hills is the only substantial mineral resource. Mostly exported for smelting in USSR at Volgograd ((28) *Why there?*) *Aluminium* ingots re-imported for fabricating, especially at Szckesfehervar, south-west of Budapest. Present plans are to increase the proportion of bauxite refined in Hungary.

Steel refining mainly at Dunaújváros (*see below*). Other widely-scattered, similar works use local ore and imported coal.

Both steel and aluminium used mainly in *vehicle manufacturing*. Ikarus works (Budapest) the biggest bus factory in Europe—output 7000 p.a., exported largely to Comecon countries but also, increasingly, to Western Europe. Other *engineering* industries: diesel engines at Gyor; heavy electrical gear, farm machinery, machine tools, telephone equipment—all mainly in and near Budapest.

Petro-chemicals—see text page 264.

Resources of industrial raw materials in South-east Europe are widely scattered and relatively scanty. There are many very small bituminous coalfields, but only those in Romania and Hungary yield sizeable outputs. There is also a great scarcity of iron ore.

To conform with Communist doctrine, however, until very recently industrial growth was largely confined to basic industries such as steel, engineering and chemicals.

South-east Europe has thus become depe
ent on imported raw materials, mainly fr
the U.S.S.R. This dependence is reflecte
the location of many modern works; e.g.
steel-works at Dunaújváros and Galati
both sited on the Danube to allow easy acc
for water-borne ore and coking coal fr
Russia. These are transhipped from ca
vessel to river barge at Brăila. (29) Dra
labelled sketch map to illustrate this po

ngineering. Oil-drilling and refinery equipment. Also (*mainly in or* *Bucharest*) diesel locomotives, tractors, machine tools and *rical* equipment.

her leading industries: *Chemicals*, based on deposits of salt and -ferrous metal ores, and on timber; *textiles*, mainly cottons; *wood-* *ing* industries—27% of Romania is forest-covered.

BULGARIA

Steel, as usual, the first modern industry to be established in the post-1947 expansion, despite serious lack of fuel and power. Steelworks at Pernik, Dimitrovo and Kremi-kovči (near Sofia) based on Russian coal and local iron ore.

The Sofia-Pernik area contains all the steel and much of the *engineering*, *textile* and *chemical* industries, and produces a quarter of Bulgaria's total manufacturing output.

Other industrial centres: Plovdiv (*food-processing, textiles, non-ferrous metallurgy*); Dimitrovgrod (*chemicals* and *electricity*, both based on local lignite); Varna—growing rapidly in importance—(*ship-building, oil-refining, textiles, chemicals*); Reka Devnja (*cement, chemicals*).

YUGOSLAVIA

Industry, as so far developed, is based almost entirely on mineral resources.

Steel—substantial output from works in area north and north-west of Sarajevo, based on locally-mined iron ore and on imported coal.

Aluminium—refined from bauxite mined near Mostar and in Montenegro.

Lead, zinc and copper—mined in the Dinaric Alps.

Oil and gas from a small field in the Danubian Plain.

Engineering, mainly in Zagreb and Belgrade, is expanding. Main products: machine tools, electrical equipment, motor vehicles (assembled from imported parts.)

Other industrial plants have been sited in irely rural areas to provide "growth tres" capable of absorbing peasants dis-ced by collectivization and modernization the farms. Examples include the steel-rks at Zenica in Yugoslavia and the chemi-works at Kędzierzin in Poland.

till more development is taking place near ated but in some cases sizeable resources petroleum, gas, non-ferrous ores and hydro-electric power. (*30*) Name six such projects from the notes above.

There has been widespread dissatisfaction with the rigid programmes of heavy indus-trial development which had little obvious effect on individual standards of living, and more light engineering products and "con-sumer goods" are now being manufactured. (*31*) Give examples from the notes above.

Transport is still rudimentary in parts of Eastern Europe. Above left: market day in Otocac, Yugoslavia. Right: Lenin Square, Sofia. Traffic densities remain lower than in most West European cities.

The main manufacturing centres are Zagreb (822 000) and the capital and river port of Belgrade (600 000). These are linked by one of Yugoslavia's few modern motor roads. The only important coastal town is Rijeka (127 000) where a recently-built refinery and chemical works relies on imported oil. (*32*) Suggest why there are very few sizeable settlements on the Yugoslav coast.

Albania has considerable natural resources, including a wide range of minerals (*see map, p. 256*) and many rivers suitable for generating power. There is, however, no good coal.

Attempts to establish manufacturing date back only to the 1920s. Most of the country's few engineering, chemical and consumer goods factories, as well as the only oil refinery, lie around Durrës and Tiranë in the centre of the Albanian plain. There are plans to build a steelworks at Elbasan, using iron ore from Pogradec and imported coal. The chief obstacle to development is Albania's antiquated transport network. Few roads are tarred and there are only 180 km of railway. The mountainous interior remains almost untouched by the 20th century.

INDEX

Where several entries are given against an item the use of heavy print indicates a major reference.

ACKNOWLEDGMENTS

We would like to thank the following for their kind permission to reproduce photographs used in this book (page references in brackets):

Aerofilms Ltd. (55, 58, 76, 111 (top and bottom), 136); Barnaby's Picture Library (188); Bavaria Verlag (98 (left and right)); Bergen Line (34); Camera Press (25 (bottom), 44 (left and right), 72, 77, 84, 85, 91, 145, 180, 195, 198, 234, 250, 253, 256 (top), 265, 268); J. Allan Cash Ltd. (21, 36, 73 (right), 123, 148-9, 166, 208, 220, 258); Danish Department of Agriculture (51); Deutschen Zentrale für Frendenverkehr, Frankfurt (7, 86 (middle and bottom), 93, 95, 102); Documentation Française (22, 60-1, 61, 70 (left and right)); Danilo Dolci Trust (148-9 (bottom)); F.C.I. (255); Fiat (153, 154); Finnish Tourist Association (20); Finnish Travel Information Centre (46); Fox Photos Ltd. (163, 175); French Ministry of Agriculture (60); French Government Tourist Office (73 (left)); Interfoto di Cardelli (157 (top)); Institut Belge d'Information et de Documentation (122 (top left and right), 124, 127); Italian State Tourist Office (159); J. H. Lowry (35); Luossavaara Kiruna (40 (top)); National Travel Association of Denmark (50-1); Notizie I.R.I. (173); Novosti Press Agency (23, 211, 212, 214, 215, 216-17, 218, 227, 229); Nowico (50); Paul Popper Ltd. (24, 82, 134, 164, 168-9, 183, 235, 236-7); Polish Cultural Institute (245); Portuguese State Office (193 (top and bottom), 194); Presse und Informationsamt der Bundesregierung, Bonn (25 (top), 86 (top), 88, 103); Enid Radcliffe (160-1, 170); Royal Netherlands Embassy (106, 108 (top left and right, middle left and right, bottom), 117); Rumanian People's Republic Legation (13, 259); Photo Simca (22); Society for Cultural Relations with U.S.S.R. (200-1, 220, 225); S.P.A.D.E.M. (68); Spanish Ministry of Information and Tourism (179); Swedish Tourist Association, Stockholm (40 (middle)); Swedish State Railways (40 (bottom)); Swiss National Tourist Office (133); Photo Yan (69).